FIPS PUB 186-4

FEDERAL INFORMATION PROCESSING STANDARDS PUBLICATION

Digital Signature Standard (DSS)

CATEGORY: COMPUTER SECURITY   SUBCATEGORY: CRYPTOGRAPHY

Information Technology Laboratory

National Institute of Standards and Technology

Gaithersburg, MD 20899-8900

Issued July 2013

U.S. Department of Commerce

Cameron F. Kerry, Acting Secretary

National Institute of Standards and Technology

Patrick D. Gallagher, Under Secretary of Commerce for Standards and Technology and Director

FOREWORD

The Federal Information Processing Standards Publication Series of the National Institute of Standards and Technology (NIST) is the official series of publications relating to standards and guidelines adopted and promulgated under the provisions of the Federal Information Security Management Act (FISMA) of 2002.

Comments concerning FIPS publications are welcomed and should be addressed to the Director, Information Technology Laboratory, National Institute of Standards and Technology, 100 Bureau Drive, Stop 8900, Gaithersburg, MD 20899-8900.

Charles Romine, Director
Information Technology Laboratory

Abstract

This Standard specifies a suite of algorithms that can be used to generate a digital signature. Digital signatures are used to detect unauthorized modifications to data and to authenticate the identity of the signatory. In addition, the recipient of signed data can use a digital signature as evidence in demonstrating to a third party that the signature was, in fact, generated by the claimed signatory. This is known as non-repudiation, since the signatory cannot easily repudiate the signature at a later time.

Key words: computer security, cryptography, digital signatures, Federal Information Processing Standards, public key cryptography.

Federal Information Processing Standards Publication 186-4

July 2013

Announcing the
DIGITAL SIGNATURE STANDARD (DSS)

Federal Information Processing Standards Publications (FIPS PUBS) are issued by the National Institute of Standards and Technology (NIST) after approval by the Secretary of Commerce pursuant to Section 5131 of the Information Technology Management Reform Act of 1996 (Public Law 104-106), and the Computer Security Act of 1987 (Public Law 100-235).

1. **Name of Standard**: Digital Signature Standard (DSS) (FIPS 186-4).

2. **Category of Standard**: Computer Security. **Subcategory.** Cryptography.

3. **Explanation**: This Standard specifies algorithms for applications requiring a digital signature, rather than a written signature. A digital signature is represented in a computer as a string of bits. A digital signature is computed using a set of rules and a set of parameters that allow the identity of the signatory and the integrity of the data to be verified. Digital signatures may be generated on both stored and transmitted data.

Signature generation uses a private key to generate a digital signature; signature verification uses a public key that corresponds to, but is not the same as, the private key. Each signatory possesses a private and public key pair. Public keys may be known by the public; private keys are kept secret. Anyone can verify the signature by employing the signatory's public key. Only the user that possesses the private key can perform signature generation.

A hash function is used in the signature generation process to obtain a condensed version of the data to be signed; the condensed version of the data is often called a message digest. The message digest is input to the digital signature algorithm to generate the digital signature. The hash functions to be used are specified in the Secure Hash Standard (SHS), FIPS 180. FIPS **approved** digital signature algorithms **shall** be used with an appropriate hash function that is specified in the SHS.

The digital signature is provided to the intended verifier along with the signed data. The verifying entity verifies the signature by using the claimed signatory's public key and the same hash function that was used to generate the signature. Similar procedures may be used to generate and verify signatures for both stored and transmitted data.

4. **Approving Authority:** Secretary of Commerce.

5. Maintenance Agency: Department of Commerce, National Institute of Standards and Technology, Information Technology Laboratory, Computer Security Division.

6. Applicability: This Standard is applicable to all Federal departments and agencies for the protection of sensitive unclassified information that is not subject to section 2315 of Title 10, United States Code, or section 3502 (2) of Title 44, United States Code. This Standard **shall** be used in designing and implementing public key-based signature systems that Federal departments and agencies operate or that are operated for them under contract. The adoption and use of this Standard is available to private and commercial organizations.

7. Applications: A digital signature algorithm allows an entity to authenticate the integrity of signed data and the identity of the signatory. The recipient of a signed message can use a digital signature as evidence in demonstrating to a third party that the signature was, in fact, generated by the claimed signatory. This is known as non-repudiation, since the signatory cannot easily repudiate the signature at a later time. A digital signature algorithm is intended for use in electronic mail, electronic funds transfer, electronic data interchange, software distribution, data storage, and other applications that require data integrity assurance and data origin authentication.

8. Implementations: A digital signature algorithm may be implemented in software, firmware, hardware or any combination thereof. NIST has developed a validation program to test implementations for conformance to the algorithms in this Standard. Information about the validation program is available at http://csrc.nist.gov/cryptval. Examples for each digital signature algorithm are available at http://csrc.nist.gov/groups/ST/toolkit/examples.html.

Agencies are advised that digital signature key pairs **shall not** be used for other purposes.

9. Other Approved Security Functions: Digital signature implementations that comply with this Standard **shall** employ cryptographic algorithms, cryptographic key generation algorithms, and key establishment techniques that have been approved for protecting Federal government sensitive information. Approved cryptographic algorithms and techniques include those that are either:

  a. specified in a Federal Information Processing Standard (FIPS),

  b. adopted in a FIPS or a NIST Recommendation, or

  c. specified in the list of approved security functions for FIPS 140.

10. Export Control: Certain cryptographic devices and technical data regarding them are subject to Federal export controls. Exports of cryptographic modules implementing this Standard and technical data regarding them must comply with these Federal regulations and be licensed by the Bureau of Industry and Security of the U.S. Department of Commerce. Information about export regulations is available at: http://www.bis.doc.gov.

11. Patents: The algorithms in this Standard may be covered by U.S. or foreign patents.

12. Implementation Schedule: This Standard becomes effective immediately upon approval by the Secretary of Commerce. A transition strategy for validating algorithms and cryptographic modules will be posted on NIST's Web page at http://csrc.nist.gov/groups/STM/cmvp/index.html under Notices. The transition plan addresses the transition by Federal agencies from modules tested and validated for compliance to previous versions of this Standard to modules tested and validated for compliance to FIPS 186-4 under the Cryptographic Module Validation Program. The transition plan allows Federal agencies and vendors to make a smooth transition to FIPS 186-4.

13. Specifications: Federal Information Processing Standard (FIPS) 186-4 Digital Signature Standard (affixed).

14. Cross Index: The following documents are referenced in this Standard. Unless a specific version or date is indicated with the document number, the latest version of the given document is intended as the reference.

    a. FIPS PUB 140, Security Requirements for Cryptographic Modules.

    b. FIPS PUB 180 Secure Hash Standard.

    c. ANS X9.31-1998, Digital Signatures Using Reversible Public Key Cryptography for the Financial Services Industry (rDSA).

    d. ANS X9.62-2005, Public Key Cryptography for the Financial Services Industry: The Elliptic Curve Digital Signature Algorithm (ECDSA).

    e. ANS X9.80, Prime Number Generation, Primality Testing and Primality Certificates.

    f. Public Key Cryptography Standard (PKCS) #1, RSA Encryption Standard.

    g. Special Publication (SP) 800-57, Recommendation for Key Management.

    h. Special Publication (SP) 800-89, Recommendation for Obtaining Assurances for Digital Signature Applications.

    i. Special Publication (SP) 800-90A, Recommendation for Random Number Generation Using Deterministic Random Bit Generators.

    j. Special Publication (SP) 800-102, Recommendation for Digital Signature Timeliness.

    k. Special Publication (SP) 800-131A, Transitions: Recommendation for Transitioning the Use of Cryptographic Algorithms and Key Lengths.

    l. IEEE Std. 1363-2000, Standard Specifications for Public Key Cryptography.

15. Qualifications: The security of a digital signature system is dependent on maintaining the secrecy of the signatory's private keys. Signatories **shall**, therefore, guard against the disclosure of their private keys. While it is the intent of this Standard to specify general security requirements for generating digital signatures, conformance to this Standard does not assure that a particular implementation is secure. It is the responsibility of an implementer to ensure that any module that implements a digital signature capability is designed and built in a secure manner.

Similarly, the use of a product containing an implementation that conforms to this Standard does not guarantee the security of the overall system in which the product is used. The responsible authority in each agency or department **shall** assure that an overall implementation provides an acceptable level of security.

Since a standard of this nature must be flexible enough to adapt to advancements and innovations in science and technology, this Standard will be reviewed every five years in order to assess its adequacy.

16. Waiver Procedure: The Federal Information Security Management Act (FISMA) does not allow for waivers to Federal Information Processing Standards (FIPS) that are made mandatory by the Secretary of Commerce.

17. Where to Obtain Copies of the Standard: This publication is available by accessing http://csrc.nist.gov/publications/. Other computer security publications are available at the same web site.

Table of Contents

Federal Information Processing Standards Publication 186-4

July 2013

Specifications for the

DIGITAL SIGNATURE STANDARD (DSS)

1. Introduction

This Standard defines methods for digital signature generation that can be used for the protection of binary data (commonly called a message), and for the verification and validation of those digital signatures. Three techniques are approved.

(1) The Digital Signature Algorithm (DSA) is specified in this Standard. The specification includes criteria for the generation of domain parameters, for the generation of public and private key pairs, and for the generation and verification of digital signatures.

(2) The RSA digital signature algorithm is specified in American National Standard (ANS) X9.31 and Public Key Cryptography Standard (PKCS) #1. FIPS 186-4 approves the use of implementations of either or both of these standards and specifies additional requirements.

(3) The Elliptic Curve Digital Signature Algorithm (ECDSA) is specified in ANS X9.62. FIPS 186-4 approves the use of ECDSA and specifies additional requirements. Recommended elliptic curves for Federal Government use are provided herein.

This Standard includes requirements for obtaining the assurances necessary for valid digital signatures. Methods for obtaining these assurances are provided in NIST Special Publication (SP) 800-89, *Recommendation for Obtaining Assurances for Digital Signature Applications*.

2. Glossary of Terms, Acronyms and Mathematical Symbols

2.1 Terms and Definitions

Approved	FIPS-approved and/or NIST-recommended. An algorithm or technique that is either 1) specified in a FIPS or NIST Recommendation, or 2) adopted in a FIPS or NIST Recommendation or 3) specified in a list of NIST approved security functions.
Assurance of domain parameter validity	Confidence that the domain parameters are arithmetically correct.
Assurance of possession	Confidence that an entity possesses a private key and any associated keying material.
Assurance of public key validity	Confidence that the public key is arithmetically correct.
Bit string	An ordered sequence of 0's and 1's. The leftmost bit is the most significant bit of the string. The rightmost bit is the least significant bit of the string.
Certificate	A set of data that uniquely identifies a key pair and an owner that is authorized to use the key pair. The certificate contains the owner's public key and possibly other information, and is digitally signed by a Certification Authority (i.e., a trusted party), thereby binding the public key to the owner.
Certification Authority (CA)	The entity in a Public Key Infrastructure (PKI) that is responsible for issuing certificates and exacting compliance with a PKI policy.
Claimed signatory	From the verifier's perspective, the claimed signatory is the entity that purportedly generated a digital signature.
Digital signature	The result of a cryptographic transformation of data that, when properly implemented, provides a mechanism for verifying origin authentication, data integrity and signatory non-repudiation.
Domain parameter seed	A string of bits that is used as input for a domain parameter generation or validation process.
Domain parameters	Parameters used with cryptographic algorithms that are usually common to a domain of users. A DSA or ECDSA cryptographic key pair is associated with a specifc set of domain parameters.

Entity	An individual (person), organization, device or process. Used interchangeably with "party".
Equivalent process	Two processes are equivalent if, when the same values are input to each process (either as input parameters or as values made available during the process or both), the same output is produced.
Hash function	A function that maps a bit string of arbitrary length to a fixed length bit string. Approved hash functions are specified in FIPS 180 and are designed to satisfy the following properties:

1. (One-way) It is computationally infeasible to find any input that maps to any new pre-specified output, and

2. (Collision resistant) It is computationally infeasible to find any two distinct inputs that map to the same output.

Hash value	See "message digest".
Intended signatory	An entity that intends to generate digital signatures in the future.
Key	A parameter used in conjunction with a cryptographic algorithm that determines its operation. Examples applicable to this Standard include:

1. The computation of a digital signature from data, and

2. The verification of a digital signature.

Key pair	A public key and its corresponding private key.
Message	The data that is signed. Also known as "signed data" during the signature verification and validation process.
Message digest	The result of applying a hash function to a message. Also known as "hash value".
Non-repudiation	A service that is used to provide assurance of the integrity and origin of data in such a way that the integrity and origin can be verified and validated by a third party as having originated from a specific entity in possession of the private key (i.e., the signatory).
Owner	A key pair owner is the entity that is authorized to use the private key of a key pair.
Party	An individual (person), organization, device or process. Used interchangeably with "entity".
Per-message secret number	A secret random number that is generated prior to the generation of each digital signature.

Public Key Infrastructure (PKI)	A framework that is established to issue, maintain and revoke public key certificates.
Prime number generation seed	A string of random bits that is used to determine a prime number with the required characteristics.
Private key	A cryptographic key that is used with an asymmetric (public key) cryptographic algorithm. For digital signatures, the private key is uniquely associated with the owner and is not made public. The private key is used to compute a digital signature that may be verified using the corresponding public key.
Probable prime	An integer that is believed to be prime, based on a probabilistic primality test. There should be no more than a negligible probability that the so-called probable prime is actually composite.
Provable prime	An integer that is either constructed to be prime or is calculated to be prime using a primality-proving algorithm.
Pseudorandom	A process or data produced by a process is said to be pseudorandom when the outcome is deterministic, yet also effectively random as long as the internal action of the process is hidden from observation. For cryptographic purposes, "effectively" means "within the limits of the intended security strength."
Public key	A cryptographic key that is used with an asymmetric (public key) cryptographic algorithm and is associated with a private key. The public key is associated with an owner and may be made public. In the case of digital signatures, the public key is used to verify a digital signature that was signed using the corresponding private key.
Security strength	A number associated with the amount of work (that is, the number of operations) that is required to break a cryptographic algorithm or system. Sometimes referred to as a security level.
Shall	Used to indicate a requirement of this Standard.
Should	Used to indicate a strong recommendation, but not a requirement of this Standard.
Signatory	The entity that generates a digital signature on data using a private key.
Signature generation	The process of using a digital signature algorithm and a private key to generate a digital signature on data.

Signature validation	The (mathematical) verification of the digital signature and obtaining the appropriate assurances (e.g., public key validity, private key possession, etc.).
Signature verification	The process of using a digital signature algorithm and a public key to verify a digital signature on data.
Signed data	The data or message upon which a digital signature has been computed. Also, see "message".
Subscriber	An entity that has applied for and received a certificate from a Certificate Authority.
Trusted third party (TTP)	An entity other than the owner and verifier that is trusted by the owner or the verifier or both. Sometimes shortened to "trusted party".
Verifier	The entity that verifies the authenticity of a digital signature using the public key.

2.2 Acronyms

ANS	American National Standard.
CA	Certification Authority.
DSA	Digital Signature Algorithm; specified in this Standard.
ECDSA	Elliptic Curve Digital Signature Algorithm; specified in ANS X9.62.
FIPS	Federal Information Processing Standard.
NIST	National Institute of Standards and Technology.
PKCS	Public Key Cryptography Standard.
PKI	Public Key Infrastructure.
RBG	Random Bit Generator.
RSA	Algorithm developed by Rivest, Shamir and Adleman; specified in ANS X9.31 and PKCS #1.
SHA	Secure Hash Algorithm; specified in FIPS 180.
SP	NIST Special Publication
TTP	Trusted Third Party.

2.3   Mathematical Symbols

$a \bmod n$	The unique remainder r, $0 \leq r \leq (n-1)$, when integer a is divided by the positive integer n. For example, $23 \bmod 7 = 2$.
$b \equiv a \bmod n$	There exists an integer k such that $b - a = kn$; equivalently, $a \bmod n = b \bmod n$.
counter	The counter value that results from the domain parameter generation process when the domain parameter seed is used to generate DSA domain parameters.
d	1. For RSA, the private signature exponent of a private key.
	2. For ECDSA, the private key.
domain_parameter_seed	A seed used for the generation of domain parameters.
e	The public verification exponent of an RSA public key.
g	One of the DSA domain parameters; g is a generator of the q-order cyclic group of $GF(p)^*$; that is, an element of order q in the multiplicative group of $GF(p)$.
GCD (a, b)	Greatest common divisor of the integers a and b.
Hash (M)	The result of a hash computation (message digest or hash value) on message M using an approved hash function.
index	A value used in the generation of g to indicate its intended use (e.g., for digital signatures).
k	For DSA and ECDSA, a per-message secret number.
L	For DSA, the length of the parameter p in bits.
(L, N)	The associated pair of length parameters for a DSA key pair, where L is the length of p, and N is the length of q.
LCM (a, b)	The least common multiple of the integers a and b.
len (a)	The length of a in bits; the integer L, where $2^{L-1} \leq a < 2^L$.
M	The message that is signed using the digital signature algorithm.
m	For ECDSA, the degree of the finite field GF_{2^m}.
N	For DSA, the length of the parameter q in bits.

n	1. For RSA, the modulus; the bit length of n is considered to be the key size.
	2. For ECDSA, the order of the base point of the elliptic curve; the bit length of n is considered to be the key size.
(n, d)	An RSA private key, where n is the modulus, and d is the private signature exponent.
(n, e)	An RSA public key, where n is the modulus, and e is the public verification exponent.
$nlen$	The length of the RSA modulus n in bits.
p	1. For DSA, one of the DSA domain parameters; a prime number that defines the Galois Field GF(p) and is used as a modulus in the operations of GF(p).
	2. For RSA, a prime factor of the modulus n.
q	1. For DSA, one of the DSA domain parameters; a prime factor of $p - 1$.
	2. For RSA, a prime factor of the modulus n.
Q	An ECDSA public key.
r	One component of a DSA or ECDSA digital signature. See the definition of (r, s).
(r, s)	A DSA or ECDSA digital signature, where r and s are the digital signature components.
s	One component of a DSA or ECDSA digital signature. See the definition of (r, s).
$seedlen$	The length of the *domain_parameter_seed* in bits.
SHA$x(M)$	The result when M is the input to the SHA-x hash function, where SHA-x is specified in FIPS 180.
x	The DSA private key.
y	The DSA public key.
$\oplus$	Bitwise logical "exclusive-or" on bit strings of the same length; for corresponding bits of each bit string, the result is determined as follows: $0 \oplus 0 = 0$, $0 \oplus 1 = 1$, $1 \oplus 0 = 1$, or $1 \oplus 1 = 0$.
	Example: $01101 \oplus 11010 = 10111$
$+$	Addition.

*	Multiplication.
/	Division.
$a \parallel b$	The concatenation of two strings a and b. Either a and b are both bit strings, or both are byte strings.
$\lceil a \rceil$	The ceiling of a: the smallest integer that is greater than or equal to a. For example, $\lceil 5 \rceil = 5$, $\lceil 5.3 \rceil = 6$, and $\lceil -2.1 \rceil = -2$.
$\lfloor a \rfloor$	The floor of a; the largest integer that is less than or equal to a. For example, $\lfloor 5 \rfloor = 5$, $\lfloor 5.3 \rfloor = 5$, and $\lfloor -2.1 \rfloor = -3$.
$\lvert a \rvert$	The absolute value of a; $\lvert a \rvert$ is $-a$ if $a < 0$; otherwise, it is simply a. For example, $\lvert 2 \rvert = 2$, and $\lvert -2 \rvert = 2$.
$[a, b]$	The interval of integers between and including a and b. For example, $[1, 4]$ consists of the integers 1, 2, 3 and 4.
$\{, a, b, \ldots\}$	Used to indicate optional information.
0x	The prefix to a bit string that is represented in hexadecimal characters.

3.   General Discussion

A digital signature is an electronic analogue of a written signature; the digital signature can be used to provide assurance that the claimed signatory signed the information. In addition, a digital signature may be used to detect whether or not the information was modified after it was signed (i.e., to detect the integrity of the signed data). These assurances may be obtained whether the data was received in a transmission or retrieved from storage. A properly implemented digital signature algorithm that meets the requirements of this Standard can provide these services.

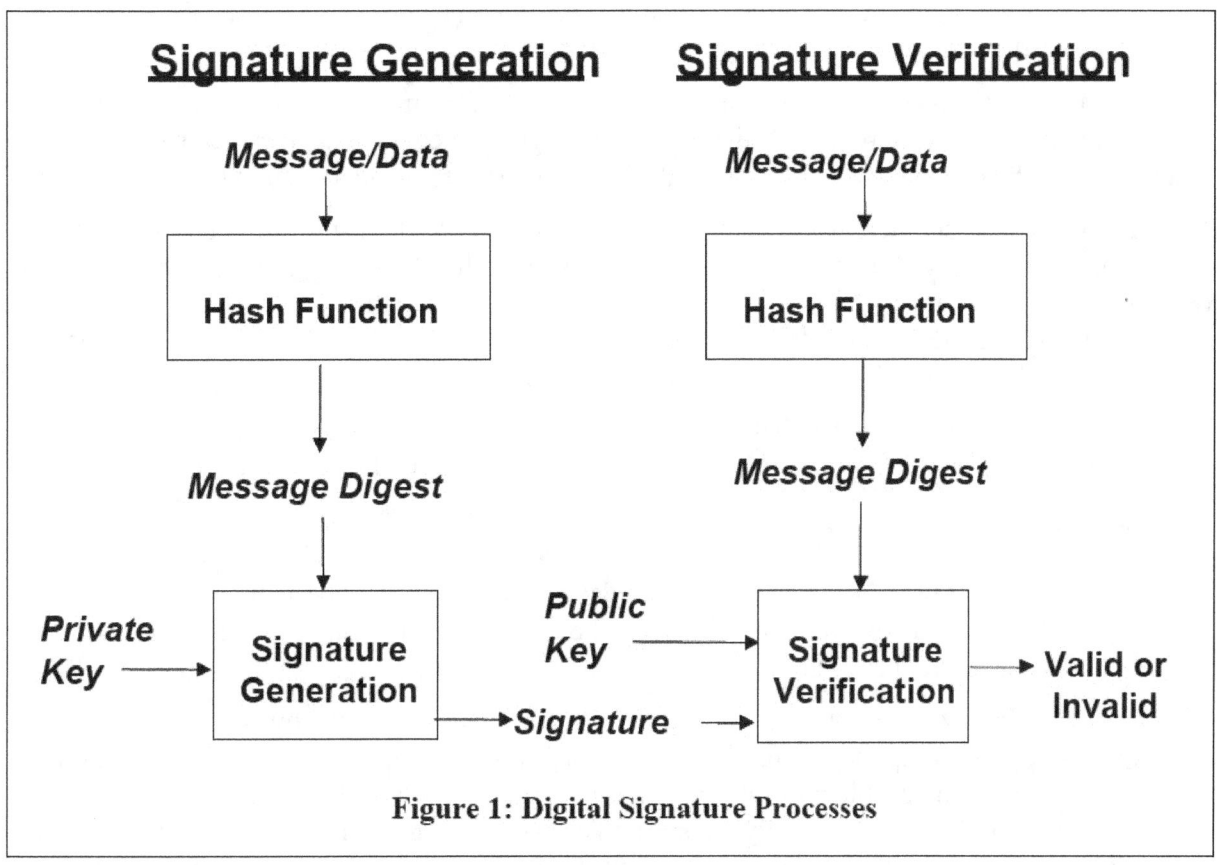

Figure 1: Digital Signature Processes

A digital signature algorithm includes a signature generation process and a signature verification process. A signatory uses the generation process to generate a digital signature on data; a verifier uses the verification process to verify the authenticity of the signature. Each signatory has a public and private key and is the owner of that key pair. As shown in Figure 1, the private key is used in the signature generation process. The key pair owner is the only entity that is authorized to use the private key to generate digital signatures. In order to prevent other entities from claiming to be the key pair owner and using the private key to generate fraudulent signatures, the private key must remain secret. The approved digital signature algorithms are designed to prevent an adversary who does not know the signatory's private key from generating the same signature as the signatory on a different message. In other words, signatures are designed so that they cannot be forged. A number of alternative terms are used in this Standard to refer to the signatory or key pair owner. An entity that intends to generate digital signatures in the future may be referred to as the *intended signatory*. Prior to the verification of a signed message, the signatory is referred to as the *claimed signatory* until such time as adequate assurance can be obtained of the actual identity of the signatory.

The public key is used in the signature verification process (see Figure 1). The public key need not be kept secret, but its integrity must be maintained. Anyone can verify a correctly signed message using the public key.

For both the signature generation and verification processes, the message (i.e., the signed data) is converted to a fixed-length representation of the message by means of an approved hash function. Both the original message and the digital signature are made available to a verifier.

A verifier requires assurance that the public key to be used to verify a signature belongs to the entity that claims to have generated a digital signature (i.e., the claimed signatory). That is, a verifier requires assurance that the signatory is the actual owner of the public/private key pair used to generate and verify a digital signature. A binding of an owner's identity and the owner's public key **shall** be effected in order to provide this assurance.

A verifier also requires assurance that the key pair owner actually possesses the private key associated with the public key, and that the public key is a mathematically correct key.

By obtaining these assurances, the verifier has assurance that if the digital signature can be correctly verified using the public key, the digital signature is valid (i.e., the key pair owner really signed the message). Digital signature validation includes both the (mathematical) verification of the digital signature and obtaining the appropriate assurances. The following are reasons why such assurances are required.

1. If a verifier does not obtain assurance that a signatory is the actual owner of the key pair whose public component is used to verify a signature, the problem of forging a signature is reduced to the problem of falsely claiming an identity. For example, anyone in possession of a mathematically consistent key pair can sign a message and claim that the signatory was the President of the United States. If a verifier does not require assurance that the President is actually the owner of the public key that is used to mathematically verify the message's signature, then successful signature verification provides assurance that the message has not been altered since it was signed, but does not provide assurance that the message came from the President (i.e., the verifier has assurance of the data's integrity, but source authentication is lacking).

2. If the public key used to verify a signature is not mathematically valid, the arguments used to establish the cryptographic strength of the signature algorithm may not apply.

The owner may not be the only party who can generate signatures that can be verified with that public key.

3. If a public key infrastructure cannot provide assurance to a verifier that the owner of a key pair has demonstrated knowledge of a private key that corresponds to the owner's public key, then it may be possible for an unscrupulous entity to have their identity (or an assumed identity) bound to a public key that is (or has been) used by another party. The unscrupulous entity may then claim to be the source of certain messages signed by that other party. Or, it may be possible that an unscrupulous entity has managed to obtain ownership of a public key that was chosen with the sole purpose of allowing for the verification of a signature on a specific message.

Technically, a key pair used by a digital signature algorithm could also be used for purposes other than digital signatures (e.g., for key establishment). However, a key pair used for digital signature generation and verification as specified in this Standard **shall not** be used for any other purpose. See SP 800-57 on Key Usage for further information.

A number of steps are required to enable a digital signature generation or verification capability in accordance with this Standard. All parties that generate digital signatures **shall** perform the initial setup process as discussed in Section 3.1. Digital signature generation **shall** be performed as discussed in Section 3.2. Digital signature verification and validation **shall** be performed as discussed in Section 3.3.

3.1   Initial Setup

Figure 2 depicts the steps that are performed prior to generating a digital

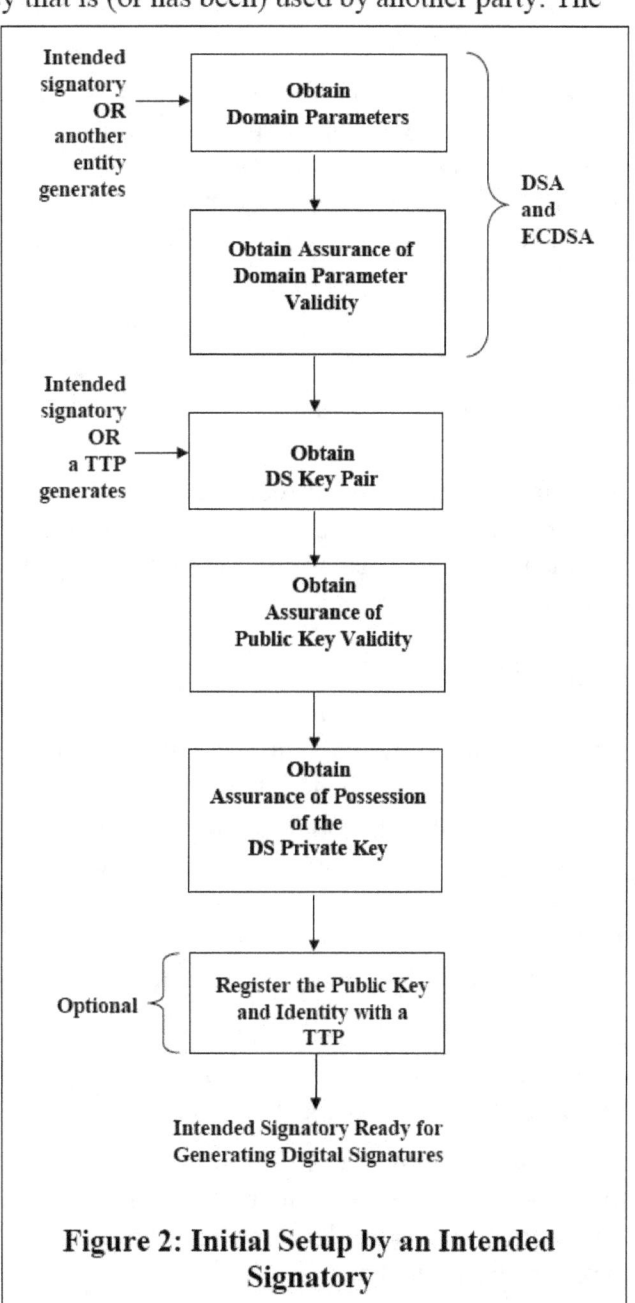

Figure 2: Initial Setup by an Intended Signatory

11

signature by an entity intending to act as a signatory.

For the DSA and ECDSA algorithms, the intended signatory **shall** first obtain appropriate domain parameters, either by generating the domain parameters itself, or by obtaining domain parameters that another entity has generated. Having obtained the set of domain parameters, the intended signatory **shall** obtain assurance of the validity of those domain parameters; approved methods for obtaining this assurance are provided in SP 800-89. Note that the RSA algorithm does not use domain parameters.

Each intended signatory **shall** obtain a digital signature key pair that is generated as specified for the appropriate digital signature algorithm, either by generating the key pair itself or by obtaining the key pair from a trusted party. The intended signatory is authorized to use the key pair and is the owner of that key pair. Note that if a trusted party generates the key pair, that party needs to be trusted not to masquerade as the owner, even though the trusted party knows the private key.

After obtaining the key pair, the intended signatory (now the key pair owner) **shall** obtain (1) assurance of the validity of the public key and (2) assurance that he/she actually possesses the associated private key. Approved methods for obtaining these assurances are provided in SP 800-89.

A digital signature verifier requires assurance of the identity of the signatory. Depending on the environment in which digital signatures are generated and verified, the key pair owner (i.e., the intended signatory) may register the public key and establish proof of identity with a mutually trusted party. For example, a certification authority (CA) could sign credentials containing an owner's public key and identity to form a certificate after being provided with proof of the owner's identity. Systems for certifying credentials and distributing certificates are beyond the scope of this Standard. Other means of establishing proof of identity (e.g., by providing identity credentials along with the public key directly to a prospective verifier) are allowed.

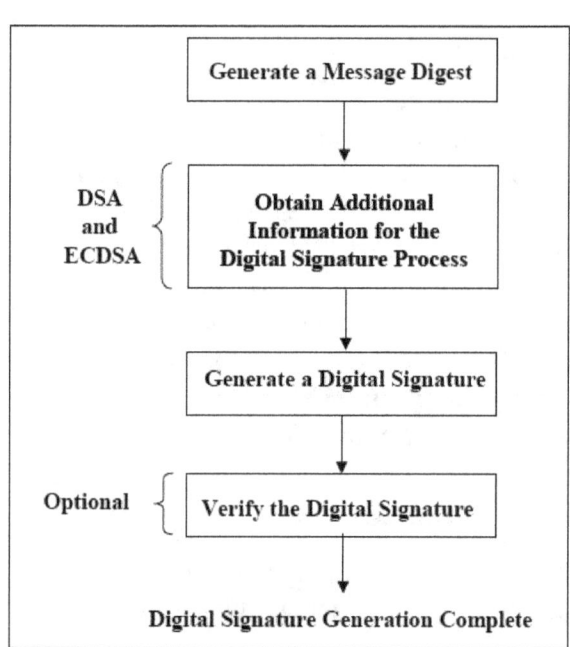

Figure 3: Digital Signature Generation

3.2 Digital Signature Generation

Figure 3 depicts the steps that are performed by an intended signatory (i.e., the entity that generates a digital signature).

Prior to the generation of a digital signature, a message digest **shall** be generated on the information to be signed using an appropriate approved hash function.

12

Depending on the digital signature algorithm to be used, additional information **shall** be obtained. For example, a random per-message secret number **shall** be obtained for DSA and ECDSA.

Using the selected digital signature algorithm, the signature private key, the message digest, and any other information required by the digital signature process, a digital signature **shall** be generated in accordance with this Standard.

The signatory may optionally verify the digital signature using the signature verification process and the associated public key. This optional verification serves as a final check to detect otherwise undetected signature generation computation errors; this verification may be prudent when signing a high-value message, when multiple users are expected to verify the signature, or if the verifier will be verifying the signature at a much later time.

3.3   Digital Signature Verification and Validation

Figure 4 depicts the digital signature verification and validation process that are performed by a verifier (e.g., the intended recipient of the signed data and associated digital signature). Note that the figure depicts a successful verification and validation process (i.e., no errors are detected).

In order to verify a digital signature, the verifier **shall** obtain the public key of the claimed signatory, (usually) based on the claimed identity. If DSA or ECDSA has been used to generate the digital signature, the verifier **shall** also obtain the domain parameters. The public key and domain parameters may be obtained, for example, from a certificate created by a trusted party (e.g., a CA) or directly from the claimed signatory. A message digest **shall** be generated on the data whose signature is to be verified (i.e., not on the received digital signature) using the same hash function that was used during the digital signature generation process. Using the appropriate digital signature algorithm, the domain parameters (if appropriate), the public key and the newly computed message digest, the received digital signature is verified in accordance with this Standard. If the verification process fails, no inference can be made as to whether the data is correct, only that in using the specified public key and the specified signature format, the digital signature cannot be verified for that data.

Before accepting the verified digital signature as valid, the verifier **shall** have (1) assurance of the signatory's claimed identity, (2) assurance of the validity of the domain parameters (for DSA and ECDSA), (3) assurance of the validity of the public key, and (4) assurance that the claimed signatory actually possessed the private key that was used to generate the digital signature at the time that the signature was generated. Methods for the verifier to obtain these assurances are provided in SP 800-89. Note that assurance of domain parameter validity may have been obtained during initial setup (see Section 3.1).

If the verification and assurance processes are successful, the digital signature and signed data **shall** be considered valid. However, if a verification or assurance process fails, the digital signature **should** be considered invalid. An organization's policy **shall** govern the action to be taken for an invalid digital signature. Such policy is outside the scope of this Standard.

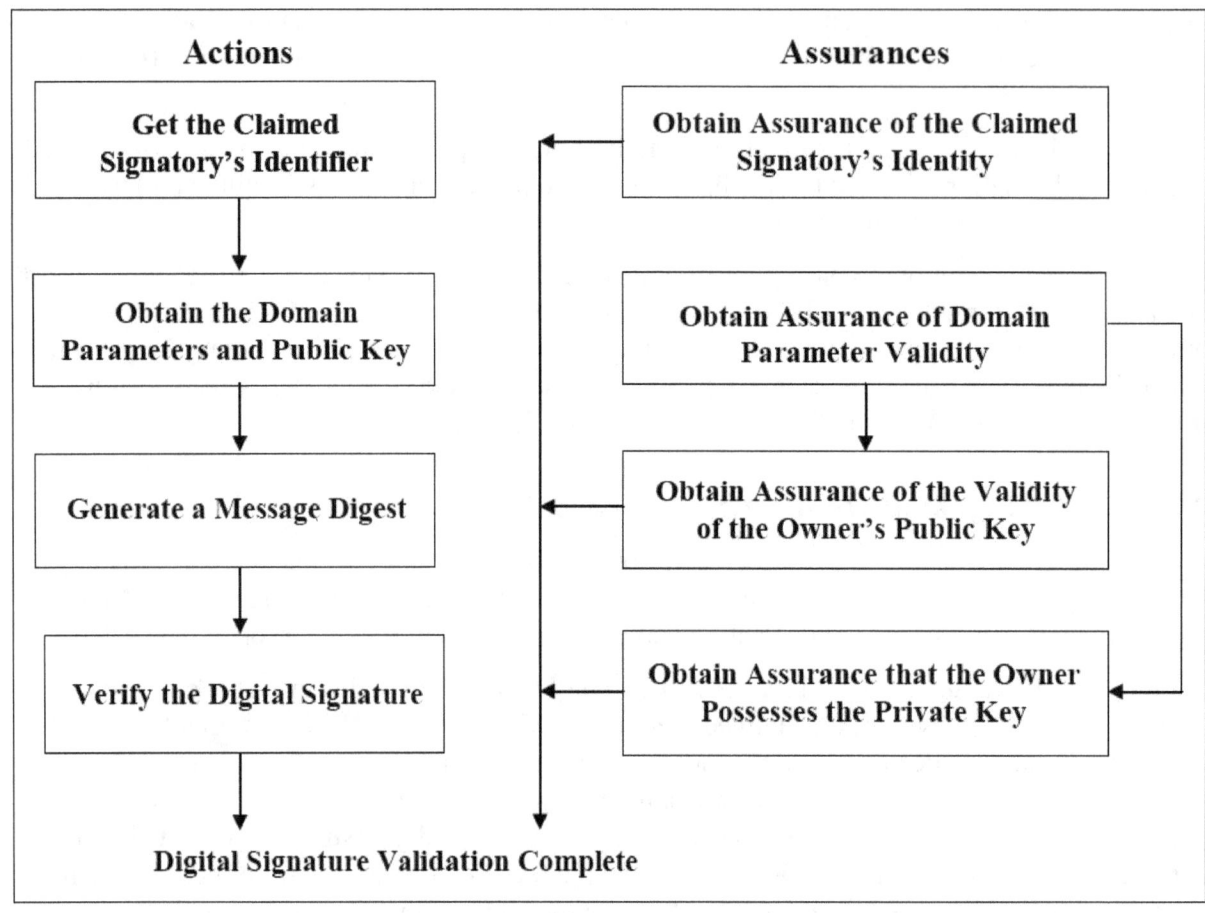

Figure 4: Digital Signature Verification and Validation

4 The Digital Signature Algorithm (DSA)

4.1 DSA Parameters

A DSA digital signature is computed using a set of domain parameters, a private key x, a per-message secret number k, data to be signed, and a hash function. A digital signature is verified using the same domain parameters, a public key y that is mathematically associated with the private key x used to generate the digital signature, data to be verified, and the same hash function that was used during signature generation. These parameters are defined as follows:

p a prime modulus, where $2^{L-1} < p < 2^L$, and L is the bit length of p. Values for L are provided in Section 4.2.

q a prime divisor of $(p - 1)$, where $2^{N-1} < q < 2^N$, and N is the bit length of q. Values for N are provided in Section 4.2.

g a generator of a subgroup of order q in the multiplicative group of GF(p), such that $1 < g < p$.

x the private key that must remain secret; x is a randomly or pseudorandomly generated integer, such that $0 < x < q$, i.e., x is in the range $[1, q-1]$.

y the public key, where $y = g^x \bmod p$.

k a secret number that is unique to each message; k is a randomly or pseudorandomly generated integer, such that $0 < k < q$, i.e., k is in the range $[1, q-1]$.

4.2 Selection of Parameter Sizes and Hash Functions for DSA

This Standard specifies the following choices for the pair L and N (the bit lengths of p and q, respectively):

$L = 1024, N = 160$

$L = 2048, N = 224$

$L = 2048, N = 256$

$L = 3072, N = 256$

Federal Government entities **shall** generate digital signatures using use one or more of these choices.

An approved hash function, as specified in FIPS 180, **shall** be used during the generation of digital signatures. The security strength associated with the DSA digital signature process is no greater than the minimum of the security strength of the (L, N) pair and the security strength of the hash function that is employed. Both the security strength of the hash function used and the security strength of the (L, N) pair **shall** meet or exceed the security strength required for the digital signature process. The security strength for each (L, N) pair and hash function is provided in SP 800-57.

SP 800-57 provides information about the selection of the appropriate (L, N) pair in accordance with a desired security strength for a given time period for the generation of digital signatures. An (L, N) pair **shall** be chosen that protects the signed information during the entire expected lifetime of that information. For example, if a digital signature is generated in 2009 for information that needs to be protected for five years, and a particular (L, N) pair is invalid after 2010, then a larger (L, N) pair **shall** be used that remains valid for the entire period of time that the information needs to be protected.

It is recommended that the security strength of the (L, N) pair and the security strength of the hash function used for the generation of digital signatures be the same unless an agreement has been made between participating entities to use a stronger hash function. When the length of the output of the hash function is greater than N (i.e., the bit length of q), then the leftmost N bits of the hash function output block **shall** be used in any calculation using the hash function output during the generation or verification of a digital signature. A hash function that provides a lower security strength than the (L, N) pair ordinarily **should not** be used, since this would reduce the security strength of the digital signature process to a level no greater than that provided by the hash function.

A Federal Government entity other than a Certification Authority (CA) **should** use only the first three (L, N) pairs (i.e., the (1024, 160), (2048, 224) and (2048, 256) pairs). A CA **shall** use an (L, N) pair that is equal to or greater than the (L, N) pairs used by its subscribers. For example, if subscribers are using the (2048, 224) pair, then the CA **shall** use either the (2048, 224), (2048, 256) or (3072, 256) pair. Possible exceptions to this rule include cross certification between CAs, certifying keys for purposes other than digital signatures and transitioning from one key size or algorithm to another. See SP 800-57 for further guidance.

4.3 DSA Domain Parameters

DSA requires that the private/public key pairs used for digital signature generation and verification be generated with respect to a particular set of domain parameters. These domain parameters may be common to a group of users and may be public. A user of a set of domain parameters (i.e., both the signatory and the verifier) **shall** have assurance of their validity prior to using them (see Section 3). Although domain parameters may be public information, they **shall** be managed so that the correct correspondence between a given key pair and its set of domain parameters is maintained for all parties that use the key pair. A set of domain parameters may remain fixed for an extended time period.

The domain parameters for DSA are the integers p, q, and g, and optionally, the *domain_parameter_seed* and *counter* that were used to generate p and q (i.e., the full set of domain parameters is $(p, q, g \{, domain\_parameter\_seed, counter\})$).

4.3.1 Domain Parameter Generation

Domain parameters may be generated by a trusted third party (a TTP, such as a CA) or by an entity other than a TTP. Assurance of domain parameter validity **shall** be obtained prior to key pair generation, digital signature generation or digital signature verification (see Section 3).

The integers p and q **shall** be generated as specified in Appendix A.1. The input to the generation process is the selected values of L and N (see Section 4.2); the output of the generation process is the values for p and q, and optionally, the values of the *domain_parameter_seed* and *counter*.

The generator g **shall** be generated as specified in Appendix A.2.

The security strength of a hash function used during the generation of the domain parameters **shall** meet or exceed the security strength associated with the (L, N) pair. Note that this is more restrictive than the hash function that can be used for the digital signature process (see Section 4.2).

4.3.2 Domain Parameter Management

Each digital signature key pair **shall** be correctly associated with one specific set of domain parameters (e.g., by a public key certificate that identifies the domain parameters associated with the public key). The domain parameters **shall** be protected from unauthorized modification until the set is deactivated (if and when the set is no longer needed). The same domain parameters may be used for more than one purpose (e.g., the same domain parameters may be used for both digital signatures and key establishment). However, using different values for the generator g reduces the risk that key pairs generated for one purpose could be accidentally used (successfully) for another purpose.

4.4 Key Pairs

Each signatory has a key pair: a private key x and a public key y that are mathematically related to each other. The private key **shall** be used for only a fixed period of time (i.e., the private key cryptoperiod) in which digital signatures may be generated; the public key may continue to be used as long as digital signatures that were generated using the associated private key need to be verified (i.e., the public key may continue to be used beyond the cryptoperiod of the associated private key). See SP 800-57 for further guidance.

4.4.1 DSA Key Pair Generation

A digital signature key pair x and y is generated for a set of domain parameters (p, q, g {, *domain_parameter_seed*, *counter*}). Methods for the generation of x and y are provided in Appendix B.1.

4.4.2 Key Pair Management

Guidance on the protection of key pairs is provided in SP 800-57. The secure use of digital signatures depends on the management of an entity's digital signature key pair as follows:

1. The validity of the domain parameters **shall** be assured prior to the generation of the key pair, or the verification and validation of a digital signature (see Section 3).

2. Each key pair **shall** be associated with the domain parameters under which the key pair was generated.

3. A key pair **shall** only be used to generate and verify signatures using the domain parameters associated with that key pair.

4. The private key **shall** be used only for signature generation as specified in this Standard and **shall** be kept secret; the public key **shall** be used only for signature verification and may be made public.

5. An intended signatory **shall** have assurance of possession of the private key prior to or concurrently with using it to generate a digital signature (see Section 3.1).

6. A private key **shall** be protected from unauthorized access, disclosure and modification.

7. A public key **shall** be protected from unauthorized modification (including substitution). For example, public key certificates that are signed by a CA may provide such protection.

8. A verifier **shall** be assured of a binding between the public key, its associated domain parameters and the key pair owner (see Section 3).

9. A verifier **shall** obtain public keys in a trusted manner (e.g., from a certificate signed by a CA that the entity trusts, or directly from the intended or claimed signatory, provided that the entity is trusted by the verifier and can be authenticated as the source of the signed information that is to be verified).

10. Verifiers **shall** be assured that the claimed signatory is the key pair owner, and that the owner possessed the private key that was used to generate the digital signature at the time that the signature was generated (i.e., the private key that is associated with the public key that will be used to verify the digital signature) (see Section 3.3).

11. A signatory and a verifier **shall** have assurance of the validity of the public key (see Sections 3.1 and 3.3).

4.5   DSA Per-Message Secret Number

A new secret random number k **shall** be generated prior to the generation of each digital signature for use during the signature generation process. This secret number **shall** be protected from unauthorized disclosure and modification.

k^{-1} is the multiplicative inverse of k with respect to multiplication modulo q; i.e., $0 < k^{-1} < q$

and $1 = (k^{-1} k) \bmod q$. This inverse is required for the signature generation process (see Section 4.6). A technique is provided in Appendix C.1 for deriving k^{-1} from k.

k and k^{-1} may be pre-computed, since knowledge of the message to be signed is not required for the computations. When k and k^{-1} are pre-computed, their confidentiality and integrity **shall** be protected.

Methods for the generation of the per-message secret number are provided in Appendix B.2.

4.6 DSA Signature Generation

The intended signatory **shall** have assurances as specified in Section 3.1.

Let N be the bit length of q. Let **min**(N, *outlen*) denote the minimum of the positive integers N and *outlen*, where *outlen* is the bit length of the hash function output block.

The signature of a message M consists of the pair of numbers r and s that is computed according to the following equations:

$r = (g^k \bmod p) \bmod q.$

$z =$ the leftmost **min**(N, *outlen*) bits of **Hash**(M).

$s = (k^{-1}(z + xr)) \bmod q.$

When computing s, the string z obtained from **Hash**(M) **shall** be converted to an integer. The conversion rule is provided in Appendix C.2.

Note that r may be computed whenever k, p, q and g are available, e.g., whenever the domain parameters p, q and g are known, and k has been pre-computed (see Section 4.5), r may also be pre-computed, since knowledge of the message to be signed is not required for the computation of r. Pre-computed k, k^{-1} and r values **shall** be protected in the same manner as the the private key x until s has been computed (see SP 800-57).

The values of r and s **shall** be checked to determine if $r = 0$ or $s = 0$. If either $r = 0$ or $s = 0$, a new value of k **shall** be generated, and the signature **shall** be recalculated. It is extremely unlikely that $r = 0$ or $s = 0$ if signatures are generated properly.

The signature (r, s) may be transmitted along with the message to the verifier.

4.7 DSA Signature Verification and Validation

Signature verification may be performed by any party (i.e., the signatory, the intended recipient or any other party) using the signatory's public key. A signatory may wish to verify that the computed signature is correct, perhaps before sending the signed message to the intended

19

recipient. The intended recipient (or any other party) verifies the signature to determine its authenticity.

Prior to verifying the signature of a signed message, the domain parameters, and the claimed signatory's public key and identity **shall** be made available to the verifier in an authenticated manner. The public key may, for example, be obtained in the form of a certificate signed by a trusted entity (e.g., a CA) or in a face-to-face meeting with the public key owner.

Let M', r', and s' be the received versions of M, r, and s, respectively; let y be the public key of the claimed signatory; and let N be the bit length of q. Also, let **min**(N, *outlen*) denote the minimum of the positive integers N and *outlen*, where *outlen* is the bit length of the hash function output block.

The signature verification process is as follows:

1. The verifier **shall** check that $0 < r' < q$ and $0 < s' < q$; if either condition is violated, the signature **shall** be rejected as invalid.

2. If the two conditions in step 1 are satisfied, the verifier computes the following:

   $w = (s')^{-1} \bmod q$.

   $z =$ the leftmost **min**(N, *outlen*) bits of **Hash**(M').

   $u1 = (zw) \bmod q$.

   $u2 = ((r')w) \bmod q$.

   $v = (((g)^{u1} (y)^{u2}) \bmod p) \bmod q$.

   A technique is provided in Appendix C.1 for deriving $(s')^{-1}$ (i.e., the multiplicative inverse of $s' \bmod q$).

   The string z obtained from **Hash**(M') **shall** be converted to an integer. The conversion rule is provided in Appendix C.2.

3. If $v = r'$, then the signature is verified. For a proof that $v = r'$ when $M' = M$, $r' = r$, and $s' = s$, see Appendix E.

4. If v does not equal r', then the message or the signature may have been modified, there may have been an error in the signatory's generation process, or an imposter (who did not know the private key associated with the public key of the claimed signatory) may have attempted to forge the signature. The signature **shall** be considered invalid. No inference can be made as to whether the data is valid, only that when using the public key to verify the signature, the signature is incorrect for that data.

5. Prior to accepting the signature as valid, the verifier **shall** have assurances as specified in Section 3.3.

An organization's policy may govern the action to be taken for invalid digital signatures. Such policy is outside the scope of this Standard. Guidance about determining the timeliness of digitally signed messages is addressed in SP 800-102, Recommendation for Digital Signature Timeliness.

5. The RSA Digital Signature Algorithm

The use of the RSA algorithm for digital signature generation and verification is specified in American National Standard (ANS) X9.31 and Public Key Cryptography Standard (PKCS) #1. While each of these standards uses the RSA algorithm, the format of the ANS X9.31 and PKCS #1 data on which the digital signature is generated differs in details that make the algorithms non-interchangeable.

5.1 RSA Key Pair Generation

An RSA digital signature key pair consists of an RSA private key, which is used to compute a digital signature, and an RSA public key, which is used to verify a digital signature. An RSA key pair used for digital signatures **shall** only be used for one digital signature scheme (e.g., ANS X9.31, RSASSA-PKCS1 v1.5 or RSASSA-PSS; see Sections 5.4 and 5.5). In addition, an RSA digital signature key pair **shall not** be used for other purposes (e.g., key establishment).

An RSA public key consists of a modulus n, which is the product of two positive prime integers p and q (i.e., $n = pq$), and a public key exponent e. Thus, the RSA public key is the pair of values (n, e) and is used to verify digital signatures. The size of an RSA key pair is commonly considered to be the length of the modulus n in bits ($nlen$).

The corresponding RSA private key consists of the same modulus n and a private key exponent d that depends on n and the public key exponent e. Thus, the RSA private key is the pair of values (n, d) and is used to generate digital signatures. Note that an alternative method for representing (n, d) using the Chinese Remainder Theorem (CRT) is allowed.

In order to provide security for the digital signature process, the two integers p and q, and the private key exponent d **shall** be kept secret. The modulus n and the public key exponent e may be made known to anyone. Guidance on the protection of these values is provided in SP 800-57.

This Standard specifies three choices for the length of the modulus (i.e., $nlen$): 1024, 2048 and 3072 bits. Federal Government entities **shall** generate digital signatures using one or more of these choices.

An approved hash function, as specified in FIPS 180, **shall** be used during the generation of key pairs and digital signatures. When used during the generation of an RSA key pair (as specified in this Standard), the length in bits of the hash function output block **shall** meet or exceed the security strength associated with the bit length of the modulus n (see SP 800-57).

The security strength associated with the RSA digital signature process is no greater than the minimum of the security strength associated with the bit length of the modulus and the security strength of the hash function that is employed. The security strength for each modulus length and hash function used during the digital signature process is provided in SP 800-57. Both the security strength of the hash function used and the security strength associated with the bit length of the modulus n **shall** meet or exceed the security strength required for the digital signature process.

It is recommended that the security strength of the modulus and the security strength of the hash function be the same unless an agreement has been made between participating entities to use a stronger hash function. A hash function that provides a lower security strength than the security strength associated with the bit length of the modulus ordinarily **should not** be used, since this would reduce the security strength of the digital signature process to a level no greater than that provided by the hash function.

Federal Government entities other than CAs **should** use only the first two choices (i.e., *nlen* = 1024 or 2048) during the timeframes indicated in SP 800-57. A CA **should** use a modulus whose length *nlen* is equal to or greater than the moduli used by its subscribers. For example, if the subscribers are using *nlen* = 2048, then the CA **should** use *nlen* ≥ 2048. SP 800-57 provides further information about the selection of the bit length of *n*. Possible exceptions to this rule include cross certification between CAs, certifying keys for purposes other than digital signatures and transitioning from one key size or algorithm to another.

Criteria for the generation of RSA key pairs are provided in Appendix B.3.1.

When RSA parameters are randomly generated (i.e., the primes p and q, and optionally, the public key exponent e), they **shall** be generated using an approved random bit generator. The (pseudo) random bits produced by the random bit generator **shall** be used as seeds for generating RSA parameters (e.g., the (pseudo) random number is used as a prime number generation seed). Prime number generation seeds **shall** be kept secret or destroyed when the modulus n is computed. If any prime number generation seed is retained (e.g., to regenerate the RSA modulus n, or as evidence that the generated prime factors p and q were generated in compliance with this Standard, then the seed **shall** be kept secret and **shall** be protected. The strength of this protection **shall** be (at least) equivalent to the protection required for the associated private key.

5.2 Key Pair Management

The secure use of digital signatures depends on the management of an entity's digital signature key pair. Key pair management requirements for RSA are provided in Section 4.4.2, requirements 4 – 11. Note that the first three requirements in Section 4.4.2, which address the relationship between domain parameters and key pairs, are not applicable to RSA.

5.3 Assurances

The intended signatory **shall** have assurances as specified in Section 3.1. Prior to accepting a digital signature as valid, the verifier **shall** have assurances as specified in Section 3.3.

5.4   ANS X9.31

ANS X9.31, *Digital Signatures Using Reversible Public Key Cryptography for the Financial Services Industry (rDSA)*, was developed for the American National Standards Institute by the Accredited Standards Committee on Financial Services, X9. See http://www.x9.org for information about obtaining copies of ANS X9.31 and any associated errata. The following discussions are based on the version of ANS X9.31 that was approved in 1998.

Methods for the generation of the private prime factors p and q are provided in Appendix B.3.

In ANS X9.31, the length of the modulus n is allowed in increments of 256 bits beyond a minimum of 1024 bits. Implementations claiming conformance to FIPS 186-4 **shall** include one or more of the modulus sizes specified in Section 5.1.

Two methods for the generation of digital signatures are included in ANS X9.31. When the public signature verification exponent e is odd, the digital signature algorithm is commonly known as RSA; when the public signature verification exponent e is even, the digital signature algorithm is commonly known as Rabin-Williams. This Standard (i.e., FIPS 186-4) adopts the use of RSA, but does not adopt the use of Rabin-Williams.

During signature verification, the extraction of the hash value $H(M)'$ from the data structure IR' **shall** be accomplished by either:

- Selecting the *hashlen* bytes of the data structure IR' that immediately precedes the two bytes of trailer information, where *hashlen* is the length in bytes of the hash function used, regardless of the length of the padding, or

- If the hash value $H(M)'$ is selected by its location with respect to the last byte of padding (i.e., 0xBA), including a check that the hash value is followed by only two bytes containing the expected trailer value.

ANS X9.31 contains an annex on random number generation. However, implementations of ANS X9.31 **shall** use approved random number generation methods.

Annexes in ANS X9.31 provide informative discussions of security and implementation considerations.

5.5   PKCS #1

Public-Key Cryptography Standard (PKCS) #1, *RSA Cryptography Standard*, defines mechanisms for encrypting and signing data using the RSA algorithm. PKCS #1 v2.1 specifies two digital signature processes and corresponding formats: RSASSA-PKCS1-v1.5 and RSASSA-PSS. Both signature schemes are approved for use, but additional constraints are imposed beyond those specified in PKCS #1 v2.1.

(a) Implementations that generate RSA key pairs **shall** use the criteria and methods in Appendix B.3 to generate those key pairs,

(b) Only approved hash functions **shall** be used.

(c) Only two prime factors p and q **shall** be used to form the modulus n.

(d) Random numbers **shall** be generated using an **approved** random bit generator.

(e) For RSASSA-PSS:

- If $nlen$ = 1024 bits (i.e., 128 bytes), and the output length of the **approved** hash function output block is 512 bits (i.e., 64 bytes), then the length (in bytes) of the salt ($sLen$) **shall** satisfy $0 \leq sLen \leq hLen - 2$,

- Otherwise, the length (in bytes) of the salt ($sLen$) **shall** satisfy $0 \leq sLen \leq hLen$,

where $hLen$ is the length of the hash function output block (in bytes).

(f) For RSASSA-PKCS-v1.5, when the hash value is recovered from the encoded message EM during the verification of the digital signature[1], the extraction of the hash value **shall** be accomplished by either:

- Selecting the rightmost (least significant) bits of EM, based on the size of the hash function used, regardless of the length of the padding, or

- If the hash value is selected by its location with respect to the last byte of padding, including a check that the hash value is located in the rightmost (least significant) bytes of EM (i.e., no other information follows the hash value in the encoded message).

Note: PKCS #1 was initially developed by RSA Laboratories in 1991 and has been revised as multiple versions. At the time of the approval of FIPS 186-4, three versions of PKSC #1 were available: version 1.5, version 2.0 and version 2.1. This Standard references only version 2.1.

---

[1] PKCS #1, v2.1 provides two methods for comparing the hash values: by comparing the encoded messages EM and EM', and by extracting the hash value from the decoding of the encoded message (see the Note in PKCS #1, v2.1). Step (f) above applies to the latter case.

6. The Elliptic Curve Digital Signature Algorithm (ECDSA)

ANS X9.62, *Public Key Cryptography for the Financial Services Industry: The Elliptic Curve Digital Signature Standard* (*ECDSA*), was developed for the American National Standards Institute by the Accredited Standards Committee on Financial Services, X9. Information about obtaining copies of ANS X9.62 is available at http://www.x9.org. The following discussions are based on the version of ANS X9.62 that was approved in 2005. This version of ANS X9.62 **shall** be used, subject to the transition period referenced in the implementation schedule of this Standard.

ANS X9.62 defines methods for digital signature generation and verification using the Elliptic Curve Digital Signature Algorithm (ECDSA). Specifications for the generation of the domain parameters used during the generation and verification of digital signatures are also included in ANS X9.62. ECDSA is the elliptic curve analog of DSA. ECDSA keys **shall not** be used for any other purpose (e.g., key establishment).

6.1   ECDSA Domain Parameters

ECDSA requires that the private/public key pairs used for digital signature generation and verification be generated with respect to a particular set of domain parameters. These domain parameters may be common to a group of users and may be public. Domain parameters may remain fixed for an extended time period.

Domain parameters for ECDSA are of the form (q, *FR*, a, b {, *domain_parameter_seed*}, G, n, h), where q is the field size; *FR* is an indication of the basis used; a and b are two field elements that define the equation of the curve; *domain_parameter_seed* is the domain parameter seed and is an optional bit string that is present if the elliptic curve was randomly generated in a verifiable fashion, G is a base point of prime order on the curve (i.e., $G = (x_G, y_G)$), n is the order of the point G, and h is the cofactor (which is equal to the order of the curve divided by n).

6.1.1   Domain Parameter Generation

This Standard specifies five ranges for n (see Table 1). For each range, a maximum cofactor size is also specified. Note that the specification of a cofactor h in a set of domain parameters is optional in ANS X9.62, whereas implementations conforming to this Standard (i.e., FIPS 186-4) **shall** specify the cofactor h in the set of domain parameters. Table 1 provides the maximum sizes for the cofactor h.

26

Table 1: ECDSA Security Parameters

Bit length of n	Maximum Cofactor (h)
160 - 223	2^{10}
224 - 255	2^{14}
256 - 383	2^{16}
384 - 511	2^{24}
≥ 512	2^{32}

ECDSA is defined for two arithmetic fields: the finite field GF_p and the finite field GF_{2^m}. For the field GF_p, p is required to be an odd prime.

NIST Recommended curves are provided in Appendix D of this Standard (i.e., FIPS 186-4). Three types of curves are provided:

1. Curves over prime fields, which are identified as P-xxx,

2. Curves over Binary fields, which are identified as B-xxx, and

3. Koblitz curves, which are identified as K-xxx,

where xxx indicates the bit length of the field size.

Alternatively, ECDSA domain parameters may be generated as specified in ANS X9.62; when ECDSA domain parameters are generated (i.e., the NIST Recommended curves are not used), the value of G **should** be generated canonically (verifiably random). An approved hash function, as specified in FIPS 180, **shall** be used during the generation of ECDSA domain parameters. When generating these domain parameters, the security strength of a hash function used **shall** meet or exceed the security strength associated with the bit length of n (see footnote 2 below).

An approved hash function, as specified in FIPS 180, is required during the generation of domain parameters. The security strength of the hash function used **shall** meet or exceed the security strength associated with the bit length of n. The security strengths for the ranges of n and the hash functions are provided in SP 800-57. It is recommended that the security strength associated with the bit length of n and the security strength of the hash function be the same

---

[2] The NIST Recommended curves were generated prior to the formulation of this guidance and using SHA-1, which was the only approved hash function available at that time. Since SHA-1 was considered secure at the time of generation, the curves were made public, and SHA-1 will only be used to validate those curves, the NIST Recommended curves are still considered secure and appropriate for Federal government use.

unless an agreement has been made between participating entities to use a stronger hash function; a hash function that provides a lower security strength than is associated with the bit length of n **shall not** be used. If the length of the output of the hash function is greater than the bit length of n, then the leftmost n bits of the hash function output block **shall** be used in any calculation using the hash function output during the generation or verification of a digital signature.

Normally, a CA **should** use a bit length of n whose assessed security strength is equal to or greater than the assessed security strength associated with the bit length of n used by its subscribers. For example, if subscribers are using a bit length of n with an assessed security strength of 112 bits, then CAs **should** use a bit length of n whose assessed security strength is equal to or greater than 112 bits. SP 800-57 provides further information about the selection of a bit length of n. Possible exceptions to this rule include cross certification between CAs, certifying keys for purposes other than digital signatures and transitioning from one key size or algorithm to another. However, these exceptions require further analysis.

6.1.2 Domain Parameter Management

Each ECDSA key pair **shall** be correctly associated with one specific set of domain parameters (e.g., by a public key certificate that identifies the domain parameters associated with the public key). The domain parameters **shall** be protected from unauthorized modification until the set is deactivated (if and when the set is no longer needed). The same domain parameters may be used for more than one purpose (e.g., the same domain parameters may be used for both digital signatures and key establishment). However, using different domain parameters reduces the risk that key pairs generated for one purpose could be accidentally used (successfully) for another purpose.

6.2   Private/Public Keys

An ECDSA key pair consists of a private key d and a public key Q that is associated with a specific set of ECDSA domain parameters; d, Q and the domain parameters are mathematically related to each other. The private key is normally used for a period of time (i.e., the cryptoperiod); the public key may continue to be used as long as digital signatures that have been generated using the associated private key need to be verified (i.e., the public key may continue to be used beyond the cryptoperiod of the associated private key). See SP 800-57 for further guidance.

ECDSA keys **shall** only be used for the generation and verification of ECDSA digital signatures.

6.2.1 Key Pair Generation

A digital signature key pair d and Q is generated for a set of domain parameters (q, FR, a, b {, *domain_parameter_seed*}, G, n, h). Methods for the generation of d and Q are provided in Appendix B.4.

6.2.2 Key Pair Management

The secure use of digital signatures depends on the management of an entity's digital signature key pair as specified in Section 4.4.2.

6.3 Secret Number Generation

A new secret random number k **shall** be generated prior to the generation of each digital signature for use during the signature generation process. This secret number **shall** be protected from unauthorized disclosure and modification. Methods for the generation of the per-message secret number are provided in Appendix B.5.

k^{-1} is the multiplicative inverse of k with respect to multiplication modulo n; i.e., $0 < k^{-1} < n$ and $1 = (k^{-1} \ k) \bmod n$. This inverse is required for the signature generation process. A technique is provided in Appendix C.1 for deriving k^{-1} from k.

k and k^{-1} may be pre-computed, since knowledge of the message to be signed is not required for the computations. When k and k^{-1} are pre-computed, their confidentiality and integrity **shall** be protected.

6.4 ECDSA Digital Signature Generation and Verification

An ECDSA digital signature (r, s) **shall** be generated as specified in ANS X9.62, using:

1. Domain parameters that are generated in accordance with Section 6.1.1,

2. A private key that is generated as specified in Section 6.2.1,

3. A per-message secret number that is generated as specified in Section 6.3,

4. An approved hash function as discussed below, and

5. An approved random bit generator.

An ECDSA digital signature **shall** be verified as specified in ANS X9.62, using the same domain parameters and hash function that were used during signature generation.

An approved hash function, as specified in FIPS 180, **shall** be used during the generation of digital signatures. The security strength associated with the ECDSA digital signature process is no greater than the minimum of the security strength associated with the bit length of n and the security strength of the hash function that is employed. Both the security strength of the hash function used and the security strength associated with the bit length of n **shall** meet or exceed the security strength required for the digital signature process. The security strengths for the ranges of the bit lengths of n and for each hash function is provided in SP 800-57.

It is recommended that the security strength associated with the bit length of n and the security strength of the hash function be the same unless an agreement has been made between participating entities to use a stronger hash function. When the length of the output of the hash function is greater than the bit length of n, then the leftmost n bits of the hash function output block **shall** be used in any calculation using the hash function output during the generation or verification of a digital signature. A hash function that provides a lower security strength than the security strength associated with the bit length of n ordinarily **should not** be used, since this would reduce the security strength of the digital signature process to a level no greater than that provided by the hash function.

6.5   Assurances

The intended signatory **shall** have assurances as specified in Section 3.1. Prior to accepting a signature as valid, the verifier **shall** have assurances as specified in Section 3.3.

APPENDIX A: Generation and Validation of FFC Domain Parameters

Finite field cryptography (FFC) is a method of implementing discrete logarithm cryptography using finite field mathematics. DSA, as specified in this Standard, is an example of FFC. The Diffie-Hellman and MQV key establishment algorithms specified in SP 800-56A can also be implemented as FFC.

The domain parameters for FFC consist of the set of values (p, q, g {, *domain_parameter_seed*, *counter*}). This appendix specifies techniques for the generation of the FFC domain parameters p, q and g and performing an explicit domain parameter validation. During the generation process, the values for *domain_parameter_seed* and *counter* are obtained.

A.1   Generation of the FFC Primes *p* and *q*

This section provides methods for generating the primes p and q that fulfill the criteria specified in Sections 4.1 and 4.2. One of these methods **shall** be used when generating these primes. A method is provided in Appendix A.1.1 to generate random candidate integers and then test them for primality using a probabilistic algorithm. A second method is provided in Appendix A.1.2 that constructs integers from smaller integers so that the constructed integer is guaranteed to be prime.

During the generation, validation and testing processes, conversions between bit strings and integers are required. Appendix C.2 provides methods for these conversions.

A.1.1   Generation and Validation of Probable Primes

Previous versions of this Standard contained a method for the generation of the domain parameters p and q using SHA-1 and probabilistic methods. This method is no longer approved for domain parameter generation; however, the validation process for this method is provided in Appendix A.1.1.1 to validate previously generated domain parameters.

A method for the generation and validation of the primes p and q using probabilistic methods is provided in Appendix A.1.1.2 and is based on the use of an approved hash function; this method **shall** be used for generating probable primes. The validation process for this method is provided in Appendix A.1.1.3.

The probabilistic methods use a hash function and an arbitrary seed (*domain_parameter_seed*). Arbitrary seeds could be anything, e.g., a user's favorite number or a random or pseudorandom number output by an approved random number generator. The *domain_parameter_seed* determines a sequence of candidates for p and q in the required intervals that are then tested for primality using a probabilistic primality test (see Appendix C.3). The test determines that the candidate is either not a prime (i.e., it is a composite integer) or is "probably a prime" (i.e., there is a very small probability that a composite integer will be declared to be a prime). p and q **shall** be the first candidate set that passes the primality tests. Note that the *domain_parameter_seed*

shall be unique for every unique set of domain parameters that are generated using the same method.

A.1.1.1 Validation of the Probable Primes *p* and *q* that were Generated Using SHA-1 as Specified in Prior Versions of this Standard

This prime validation algorithm is used to validate that the primes *p* and *q* that were generated by the prime generation algorithm specified in previous versions of this Standard. The algorithm requires the values of *p*, *q*, *domain_parameter_seed* and *counter*, which were output from the prime generation algorithm.

Let **SHA1( )** be the SHA-1 hash function specified in FIPS 180. The following process or its equivalent **shall** be used to validate *p* and *q* for this method.

Input:

    1. *p, q*   The generated primes *p* and *q*.

    2. *domain_parameter_seed* A seed that was used to generate *p* and *q*.

    3. *counter*   A count value that was determined during generation.

Output:

    1. *status*   The status returned from the validation procedure, where status is either **VALID** or **INVALID**.

Process:

    1. If (**len** (p) $\neq$ 1024) or (**len** (q) $\neq$ 160), then return **INVALID**.

    2. If (*counter* > 4095), then return **INVALID**.

    3. *seedlen* = **len** (*domain_parameter_seed*).

    4. If (*seedlen* < 160), then return **INVALID**.

    5. *computed_q* = **SHA1**(*domain_parameter_seed*) $\oplus$ **SHA1**((*domain_parameter_seed* + 1) mod $2^{seedlen}$).

    6. Set the first and last bits of *computed_q* equal to 1 (i.e., the 159th and 0th bits).

    7. Test whether or not *computed_q* is prime as specified in Appendix C.3. If (*computed_q* $\neq$ *q*) or (*computed_q* is not prime), then return **INVALID**.

    8. *offset* = 2.

    9. For *i* = 0 to *counter* do

        9.1   For *j* = 0 to 6 do

             V_j = **SHA1**((*domain_parameter_seed* + *offset* + *j*) mod $2^{seedlen}$).

        9.2 $W = V_0 + (V_1 * 2^{160}) + (V_2 * 2^{320}) + (V_3 * 2^{480}) + (V_4 * 2^{640}) + (V_5 * 2^{800}) +$

$$((V_6 \bmod 2^{63}) * 2^{960}).$$

9.3 $X = W + 2^{1023}.$ Comment: $0 \le W < 2^{L-1}$.

9.4 $c = X \bmod 2q.$

9.5 $computed\_p = X - (c - 1).$Comment: $computed\_p \equiv 1 \pmod{2q}$.

9.6   If ($computed\_p < 2^{1023}$), then go to step 9.8.

9.7   Test whether or not $computed\_p$ is prime as specified in Appendix C.3. If $computed\_p$ is determined to be prime, then go to step 10.

9.8 $offset = offset + 7.$

10. If (($i \ne counter$) or ($computed\_p \ne p$) or ($computed\_p$ is not prime)), then return **INVALID.**

11. Return **VALID.**

A.1.1.2   Generation of the Probable Primes *p* and *q* Using an Approved Hash Function

This method uses an approved hash function and may be used for the generation of the primes *p* and *q* for any application (e.g., digital signatures or key establishment). The security strength of the hash function **shall** be equal to or greater than the security strength associated with the (*L*, *N*) pair.

An arbitrary *domain_parameter_seed* of *seedlen* bits is also used, where *seedlen* **shall** be equal to or greater than *N*.

The generation process returns a set of integers *p* and *q* that have a very high probability of being prime. For another entity to validate that the primes were generated correctly using the validation process in Appendix A.1.1.3, the value of the *domain_parameter_seed* and the *counter* used to generate the primes must also be returned and made available to the validating entity; the *domain_parameter_seed* and *counter* need not be kept secret. Let **Hash( )** be the selected hash function, and let *outlen* be the bit length of the output block, where *outlen* **shall** be equal to or greater than *N*.

The following process or its equivalent **shall** be used to generate *p* and *q* for this method.

Input:

1.	*L*	The desired length of the prime *p* (in bits).
2.	*N*	The desired length of the prime *q* (in bits).
3.	*seedlen*	The desired length of the domain parameter seed; *seedlen* **shall** be equal to or greater than *N*.

Output:

1.	*status*	The status returned from the generation procedure, where status is

either **VALID** or **INVALID**. If **INVALID** is returned as the *status*, either no values for the other output parameters **shall** be returned, or invalid values **shall** be returned (e.g., zeros or Null strings).

2. *p, q* The generated primes *p* and *q*.

3. *domain_parameter_seed*

   (Optional) A seed that was used to generate *p* and *q*.

4. *counter* (Optional) A count value that was determined during generation.

Process:

1. Check that the (*L*, *N*) pair is in the list of acceptable (*L*, *N* pairs) (see Section 4.2). If the pair is not in the list, then return **INVALID.**

2. If (*seedlen* < *N*), then return **INVALID.**

3. $n = \lceil L / outlen \rceil - 1$.

4. $b = L - 1 - (n * outlen)$.

5. Get an arbitrary sequence of *seedlen* bits as the *domain_parameter_seed*.

6. $U = \textbf{Hash}\,(domain\_parameter\_seed) \bmod 2^{N-1}$.

7. $q = 2^{N-1} + U + 1 - (U \bmod 2)$.

8. Test whether or not *q* is prime as specified in Appendix C.3.

9. If *q* is not a prime, then go to step 5.

10. *offset* = 1.

11. For *counter* = 0 to (4*L* − 1) do

    11.1 For *j* = 0 to *n* do

    $V_j = \textbf{Hash}\,((domain\_parameter\_seed + offset + j) \bmod 2^{seedlen})$.

    11.2 $W = V_0 + (V_1 * 2^{outlen}) + \ldots + (V_{n-1} * 2^{(n-1)*outlen}) + ((V_n \bmod 2^b) * 2^{n*outlen})$.

    11.3 $X = W + 2^{L-1}$. Comment: $0 \le W < 2^{L-1}$; hence, $2^{L-1} \le X < 2^L$.

    11.4 $c = X \bmod 2q$.

    11.5 $p = X - (c - 1)$. Comment: $p \equiv 1 \pmod{2q}$.

    11.6 If ($p < 2^{L-1}$), then go to step 11.9.

    11.7 Test whether or not *p* is prime as specified in Appendix C.3.

    11.8 If *p* is determined to be prime, then return **VALID** and the values of *p*, *q* and (optionally) the values of *domain_parameter_seed and counter.*

34

11.9   *offset = offset + n + 1.*   Comment: Increment *offset*; then, as part of the loop in step 11, increment *counter*; if *counter* < 4L, repeat steps 11.1 through 11.8.

12. Go to step 5.

A.1.1.3   Validation of the Probable Primes *p* and *q* that were Generated Using an Approved Hash Function

This prime validation algorithm is used to validate that the integers *p* and *q* were generated by the prime generation algorithm given in Appendix A.1.1.2. The validation algorithm requires the values of *p*, *q*, *domain_parameter_seed* and *counter*, which were output from the prime generation algorithm. Let **Hash( )** be the hash function used to generate *p* and *q*, and let *outlen* be its output block length.

The following process or its equivalent **shall** be used to validate *p* and *q* for this method.

Input:

    1. *p, q*   The generated primes *p* and *q*.

    3. *domain_parameter_seed*   The domain parameter seed that was used to generate *p* and *q*.

    4. *counter*   A count value that was determined during generation.

Output:

    1. *status*   The status returned from the validation procedure, where status is either **VALID** or **INVALID**.

Process:

1. $L = \textbf{len}(p)$.

2. $N = \textbf{len}(q)$.

3. Check that the (L, N) pair is in the list of acceptable (L, N) pairs (see Section 4.2). If the pair is not in the list, return **INVALID**.

4. If $(counter > (4L - 1))$, then return **INVALID**.

5. $seedlen = \textbf{len}(domain\_parameter\_seed)$.

6. If $(seedlen < N)$, then return **INVALID**.

7. $U = \textbf{Hash}(domain\_parameter\_seed) \bmod 2^{N-1}$.

8. $computed\_q = 2^{N-1} + U + 1 - (U \bmod 2)$.

9. Test whether or not *computed_q* is prime as specified in Appendix C.3. If (*computed_q* ≠ *q*) or (*computed_q* is not prime), then return **INVALID**.

35

10. $n = \lceil L/outlen \rceil - 1$.

11. $b = L - 1 - (n * outlen)$.

12. $offset = 1$.

13. For $i = 0$ to $counter$ do

   13.1 For $j = 0$ to n do

   $$V_j = \textbf{Hash}((domain\_parameter\_seed + offset + j) \bmod 2^{seedlen}).$$

   13.2 $W = V_0 + (V_1 * 2^{outlen}) + \ldots + (V_{n-1} * 2^{(n-1)\,*\,outlen}) + ((V_n \bmod 2^b) * 2^{n\,*\,outlen})$.

   13.3 $X = W + 2^{L-1}$.

   13.4 $c = X \bmod 2q$.

   13.5 $computed\_p = X - (c - 1)$.

   13.6 If $(computed\_p < 2^{L-1})$, then go to step 13.9

   13.7 Test whether or not $computed\_p$ is prime as specified in Appendix C.3.

   13.8 If $computed\_p$ is determined to be a prime, then go to step 14.

   13.9 $offset = offset + n + 1$.

14. If $((i \neq counter)$ or $(computed\_p \neq p)$ or $(computed\_p$ is not a prime$))$, then return **INVALID**.

15. Return **VALID**.

A.1.2 Construction and Validation of the Provable Primes *p* and *q*

Primes can be generated so that they are guaranteed to be prime. The following algorithm for generating *p* and *q* uses an approved hash function and an arbitrary seed (*firstseed*) to construct *p* and *q* in the required intervals. The security strength of the hash function **shall** be equal to or greater than the security strength associated with the (*L*, *N*) pair.

Arbitrary seeds can be anything, e.g., a user's favorite number or a random or pseudorandom number that is output from a random number generator. Note that the *firstseed* must be unique to produce a unique set of domain parameters. Candidate primes are tested for primality using a deterministic primality test that proves whether or not the candidate is prime. The resulting *p* and *q* are guaranteed to be primes.

A.1.2.1 Construction of the Primes *p* and *q* Using the Shawe-Taylor Algorithm

For each set of domain parameters generated, an arbitrary initial seed (*firstseed*) of at least *seedlen* bits **shall** be determined, where *seedlen* **shall** be ≥ *N*.

The generation process returns a set of integers *p* and *q* that are guaranteed to be prime. For another entity to validate that the primes were generated correctly (using the validation process in Appendix A.1.2.2), the value of the *firstseed*, the two computed seeds (*pseed* and *qseed*) and the counters used to generate the primes (*pgen_counter* and *qgen_counter*) must be made available to the validating entity; the seeds and the counters need not be kept secret. The domain parameters for DSA are identified in Section 4.3 as (p, q, g {, *domain_parameter_seed*, *counter*}). When using the Shawe-Taylor algorithm for generating p and q, the *domain_parameter_seed* consists of three seed values (*firstseed, pseed,* and *qseed*), and the *counter* consists of the pair of counter values (*pgen_counter* and *qgen_counter*).

Let **Hash( )** be the selected hash function (see Appendix A.1.2), and let *outlen* be the bit length of the output block of that hash function.

A.1.2.1.1 Get the First Seed

The following process or its equivalent **shall** be used to generate *firstseed* for this constructive method.

Input:

1.	N	The length of q in bits.
2.	*seedlen*	The length of firstseed, where *seedlen* $\geq N$.

Output:

1.	*status*	The status returned from the generation procedure, where *status* is either **SUCCESS** or **FAILURE. If FAILURE** is returned, then either no *firstseed* value **shall** be provided or an invalid value **shall** be returned.
2.	*firstseed*	The first seed generated.

Process:

1. *firstseed* = 0.

2. Check that N is in the list of acceptable (L, N) pairs (see Section 4.2). If not, then return **FAILURE.**

3. If (*seedlen* < N), then return **FAILURE.**

4. While *firstseed* < 2^{N-1}.

   Get an arbitrary sequence of *seedlen* bits as *firstseed*.

5. Return **SUCCESS** and the value of *firstseed*.

Note: This routine could be incorporated into the beginning of the constructive prime generation procedure in Appendix A.1.2.1.2. However, this was not done in this specification so that the validation process in Appendix A.1.2.2 could also call the constructive prime generation procedure and provide the value of *firstseed* as input.

A.1.2.1.2 Constructive Prime Generation

The following process or its equivalent **shall** be used to generate p and q for this constructive method.

Input:

1.	L	The requested length for p (in bits).
2.	N	The requested length for q (in bits).
3.	*firstseed*	The first seed to be used. This was obtained as specified in Appendix A.1.2.1.1.

Output:

1.	*status*	The status returned from the generation procedure, where *status* is either **SUCCESS** or **FAILURE**. If **FAILURE** is returned, then either no other values **shall** be returned, or invalid values **shall** be returned.
2.	p, q	The requested primes.
3.	*pseed*, *qseed*	(Optional) Computed seed values that were used to generate p and q. The entire *seed* for the generation of p and q consists of *firstseed*, *pseed* and *qseed*.
4.	*pgen_counter*, *qgen_counter*	(Optional) The count values that were determined during generation.

Process:

1. Check that the (L, N) pair is in the list of acceptable (L, N) pairs (see Section 4.2). If the pair is not in the list, return **FAILURE**.

    > Comment: Use the Shawe-Taylor random prime routine in Appendix C.6 to generate random primes.

2. Using N as the length and *firstseed* as the *input_seed*, use the random prime generation routine in Appendix C.6 to obtain q, *qseed* and *qgen_counter*. If **FAILURE** is returned, then return **FAILURE**.

3. Using $\lceil L / 2 + 1 \rceil$ as the *length* and *qseed* as the *input_seed*, use the random prime routine in Appendix C.6 to obtain p_0, *pseed*, and *pgen_counter*. If **FAILURE** is returned, then return **FAILURE**.

38

4. *iterations* $= \lceil L / outlen \rceil - 1$.

5. *old_counter* = *pgen_counter*.

Comment: Generate a (pseudo) random x in the interval $[2^{L-1}, 2^L]$.

6. $x = 0$.

7. For $i = 0$ to *iterations* do

$x = x + (\textbf{Hash}(pseed + i) * 2^{i * outlen})$.

8. *pseed* = *pseed* + *iterations* + 1.

9. $x = 2^{L-1} + (x \bmod 2^{L-1})$.

Comment: Generate p, a candidate for the prime, in the interval $[2^{L-1}, 2^L]$.

10. $t = \lceil x / (2q \, p_0) \rceil$.

11. If $(2tq \, p_0 + 1) > 2^L$, then $t = \lceil 2^{L-1} / (2q \, p_0) \rceil$.

12. $p = 2tq \, p_0 + 1$.

13. *pgen_counter* = *pgen_counter* + 1.

Comment: Test p for primality; choose an integer a in the interval $[2, p-2]$.

14. $a = 0$

15. For $i = 0$ to *iterations* do

$a = a + (\textbf{Hash}(pseed + i) * 2^{i * outlen})$.

16. *pseed* = *pseed* + *iterations* + 1.

17. $a = 2 + (a \bmod (p-3))$.

18. $z = a^{2tq} \bmod p$.

19. If $((1 = \textbf{GCD}(z-1, p))$ and $(1 = z^{p_0} \bmod p))$, then return **SUCCESS** and the values of p, q and (optionally) *pseed, qseed, pgen_counter,* and *qgen_counter*.

20. If $(pgen\_counter > (4L + old\_counter))$, then **return FAILURE.**

21. $t = t + 1$.

22. Go to step 11.

A.1.2.2 Validation of the DSA Primes *p* and *q* that were Constructed Using the Shawe-Taylor Algorithm

The validation of the primes *p* and *q* that were generated by the method described in Appendix A.1.2.1.2 may be performed if the values of *firstseed*, *pseed*, *qseed*, *pgen_counter* and *qgen_counter* were saved and are provided for use in the following algorithm.

The following process or its equivalent **shall** be used to validate *p* and *q* for this constructive method.

Input:

1. *p, q*   The primes to be validated.

2. *firstseed, pseed, qseed*   Seed values that were used to generate *p* and *q*.

3. *pgen_counter, qgen_counter*

                        The count values that were determined during generation.

Output:

1. *status*   The status returned from the validation procedure, where *status* is either **SUCCESS** or **FAILURE.**

Process:

1. $L = $ **len** (p).

2. $N = $ **len** (q).

3. Check that the (L, N) pair is in the list of acceptable (L, N) pairs (see Section 4.2). If the pair is not in the list, then return **FAILURE.**

4. If $(firstseed < 2^{N-1})$, then return **FAILURE.**

5. If $(2^N \le q)$, then return **FAILURE**).

6. If $(2^L \le p)$, then return **FAILURE.**

7. If $((p - 1) \bmod q \ne 0)$, then return **FAILURE.**

8. Using L, N and *firstseed*, perform the constructive prime generation procedure in Appendix A.1.2.1.2 to obtain *p_val*, *q_val*, *pseed_val*, *qseed_val*, *pgen_counter_val*, and *qgen_counter_val*. If **FAILURE** is returned, or if $(q\_val \ne q)$ or $(qseed\_val \ne$

qseed) or (*qgen_counter_val* ≠ *qgen_counter*) or (*p_val* ≠ *p*) or (*pseed_val* ≠ *pseed*) or (*pgen_counter_val* ≠ *pgen_counter*), then return **FAILURE.**

9. Return **SUCCESS.**

A.2   Generation of the Generator *g*

The generator *g* depends on the values of *p* and *q*. Two methods for determining the generator *g* are provided; one of these methods **shall** be used. The first method, discussed in Appendix A.2.1, may be used when complete validation of the generator *g* is not required; it is recommended that this method be used only when the party generating *g* is trusted to not deliberately generate a *g* that has a potentially exploitable relationship to another generator *g'*. For example, it must be hard to determine an exponential relationship between the generators such that $g = (g')^x \mod p$ for a known value of *x*. (Note: Read $(g')^x$ as *g* prime to the *x*.)

Appendix A.2.2 provides a method for partial validation when the method of generation in Appendix A.2.1 is used. The second method for generating *g*, discussed in Appendix A.2.3, **shall** be used when validation of the generator *g* is required; the method for the validation of a generator determined using the method in Appendix A.2.3 is provided in Appendix A.2.4.

A.2.1 Unverifiable Generation of the Generator *g*

This method is used to determine a value for *g*, based on the values of *p* and *q*. It may be used when validation of the generator *g* is not required. The correct generation of *g* cannot be completely validated (see Appendix A.2.2). Note that this generation method for *g* was also specified in previous versions of this Standard.

The following process or its equivalent **shall** be used to generate the generator *g* for this method.

Input:

1. *p, q*   The generated primes.

Output:

1. *g*   The requested value of *g*.

Process:

1. $e = (p - 1)/q$.

2. Set *h* = any integer satisfying $1 < h < (p - 1)$, such that *h* differs from any value previously tried. Note that *h* could be obtained from a random number generator or from a counter that changes after each use.

3. $g = h^e \mod p$.

4. If ($g = 1$), then go to step 2.

5. Return g.

A.2.2 Assurance of the Validity of the Generator *g*

The order of the generator g that was generated using Appendix A.2.1 can be partially validated by checking the range and order, thereby performing a partial validation of g.

The following process or its equivalent **shall** be used when partial validation of the generator g is required:

Input:

    1. *p, q, g*   The domain parameters.

Output:

    1. *status*   The status returned from the generation routine, where *status* is either **PARTIALLY VALID or INVALID.**

Process:

    1. Verify that $2 \le g \le (p{-}1)$. If not true, return **INVALID.**

    2. If ($g^q = 1 \bmod p$), then return **PARTIALLY VALID.**

    3. Return **INVALID.**

The non-existence of a potentially exploitable relationship of g to another generator g' (that is known to the entity that generated g, but may not be known by other entities) cannot be checked. In this sense, the correct generation of g cannot be completely validated.

A.2.3 Verifiable Canonical Generation of the Generator *g*

The generation of g is based on the values of p, q and *domain_parameter_seed* (which are outputs of the generation processes in Appendix A.1). When p and q were generated using the method in Appendix A.1.1.2, the *domain_parameter_seed* value must have been returned from the generation routine. When p and q were generated using the method in Appendix A.1.2.1, the *firstseed*, *pseed*, and *qseed* values must have been returned from the generation routine; in this case, *domain_parameter_seed = firstseed || pseed || qseed* **shall** be used in the following process.

This method of generating a generator g can be validated (see Appendix A.2.4).

This generation method supports the generation of multiple values of g for specific values of p and q. The use of different values of g for the same p and q may be used to support key separation; for example, using the g that is generated with *index* = 1 for digital signatures and with *index* = 2 for key establishment.

Let **Hash( )** be the hash function used to generate p and q (see Appendix A.1). The following process or its equivalent **shall** be used to generate the generator g.

Input:

1. p, q The primes.

2. *domain_parameter_seed*   The seed used during the generation of p and q.

3. *index*   The index to be used for generating g. *index* is a bit string of length 8 that represents an unsigned integer.

Output:

1. *status*   The status returned from the generation routine, where *status* is either **VALID** or **INVALID.**

2. g The value of g that was generated.

Process: Note: *count* is an unsigned 16-bit integer.

                                        Comment: Check that a valid value of the *index* has been provided (see above).

1. If (*index* is incorrect), then return **INVALID.**

2. $N = $ **len**(q).

3. $e = (p - 1)/q$.

4. *count* = 0.

5. *count* = *count* + 1.

                                         Comment: Check that *count* does not wrap around to 0.

6. If (*count* = 0), then return **INVALID.**

                                         Comment: the length of the *domain_parameter_seed* has already been checked. "ggen" is the bit string 0x6767656E.

7. $U = $ *domain_parameter_seed* || "ggen" || *index* || *count*.

8. $W = $ **Hash**(U).

9. $g = W^e \bmod p$.

10. If ($g < 2$), then go to step 5.   Comment: If a generator has not been found.

11. Return **VALID** and the value of g.

A.2.4 Validation Routine when the Canonical Generation of the Generator *g* Routine Was Used

This algorithm **shall** be used to validate the value of *g* that was generated using the process in Appendix A.2.3, based on the values of *p, q, domain_parameter_seed,* and the appropriate value of *index.* It is assumed that the values of *p* and *q* have been previously validated according to Appendix A.1. Note that the method specified in Appendix A.2.3 for the generation of *g* was not included in previous versions of this Standard; therefore, this validation method is not appropriate for that case.

The *domain_parameter_seed* is an output from the generation of *p* and *q*. When *p* and *q* were generated using the method in Appendix A.1.1.2, the *domain_parameter_seed* must have been returned from the generation routine and made available to the validating party. When *p* and *q* were generated using the method in Appendix A.1.2.1, the *firstseed, pseed,* and *qseed* values must have been returned from the generation routine and made available; *firstseed, pseed,* and *qseed* **shall** be concatenated to form the *domain_parameter_seed* used in the following process. Let **Hash( )** be the hash function used to generate *g* (i.e., the hash function also used to generate *p* and *q*).

The input *index* is the index number for the generator *g*. See Appendix A.2.3 for more details.

The following process or its equivalent **shall** be used to validate the generator *g* for this method.

Input:

1.	*p, q*	The primes.
2.	*domain_parameter_seed*	The seed used to generate *p* and q.
3.	*index*	The index used in Appendix A.2.3 to generate *x. index* is a bit string of length 8 that represents an unsigned integer.
4.	*g*	The value of *g* to be validated.

Output:

1.	*status*	The status returned from the generation routine, where *status* is either **VALID** or **INVALID**.

Process:
Note: *count* is an unsigned 16-bit integer.

Comment: Check that a valid value of the *index* has been provided (see above).

1. If (*index* is incorrect), then return **INVALID.**

2. Verify that $2 \leq g \leq (p-1)$. If not true, return **INVALID**.

3. If ($g^q \neq 1 \mod p$), then return **INVALID.**

4. $N = \textbf{len}(q)$.

5. $e = (p - 1)/q$.

6. $count = 0$.

7. $count = count + 1$.

   Comment: Check that $count$ does not wrap around to 0.

8. If $(count = 0)$, then return **INVALID**.

   Comment: "ggen" is the bit string 0x6767656E.

9. $U = domain\_parameter\_seed \parallel$ "ggen" $\parallel index \parallel count$.

10. $W = \textbf{Hash}(U)$.

11. $computed\_g = W^e \bmod p$.

12. If $(computed\_g < 2)$, then go to step 7.   Comment: If a generator has not been found.

13. If $(computed\_g = g)$, then return **VALID**, else return **INVALID**.

APPENDIX B: Key Pair Generation

Discrete logarithm cryptography (DLC) is divided into finite field cryptography (FFC) and elliptic curve cryptography (ECC); the difference between the two is the type of math that is used. DSA is an example of FFC; ECDSA is an example of ECC. Other examples of DLC are the Diffie-Hellman and MQV key agreement algorithms, which have both FFC and ECC forms.

The most common example of integer factorization cryptography (IFC) is RSA.

This appendix specifies methods for the generation of FFC and ECC key pairs and secret numbers, and the generation of IFC key pairs. All generation methods require the use of an approved, properly instantiated random bit generator (RBG); the RBG **shall** have a security strength equal to or greater than the security strength associated with the key pairs and secret numbers to be generated. See SP 800-57 for guidance on security strengths and key sizes.

This appendix does not indicate the required conversions between bit strings and integers. When required by a process in this appendix, the conversion **shall** be accomplished as specified in Appendix C.2.

B.1   FFC Key Pair Generation

An FFC key pair (x, y) is generated for a set of domain parameters $(p, q, g$ {, *domain_parameter_seed, counter*}). Two methods are provided for the generation of the FFC private key x and public key y; one of these two methods **shall** be used. Prior to generating DSA key pairs, assurance of the validity of the domain parameters $(p, q$ and $g)$ **shall** have been obtained as specified in Section 3.1.

For DSA, the valid values of L and N are provided in Section 4.2.

B.1.1 Key Pair Generation Using Extra Random Bits

In this method, 64 more bits are requested from the RBG than are needed for x so that bias produced by the mod function in step 6 is negligible.

The following process or its equivalent may be used to generate an FFC key pair.

Input:

(p, q, g) The subset of the domain parameters that are used for this process. p, q and g **shall** either be provided as integers during input, or **shall** be converted to integers prior to use.

Output:

1. *status*   The status returned from the key pair generation process. The status will indicate **SUCCESS** or an **ERROR**.

2. (x, y) The generated private and public keys. If an error is encountered during the generation process, invalid values for x and y **should** be returned, as represented by *Invalid_x* and *Invalid_y* in the following specification. x and y are returned as integers. The generated private key x is in the range $[1, q–1]$, and the public key is in the range $[1, p–1]$.

Process:

1. $N = \mathbf{len}(q); L = \mathbf{len}(p)$.

> Comment: Check that the (L, N) pair is specified in Section 4.2.

2. If the (L, N) pair is invalid, then return an **ERROR** indicator, *Invalid_x*, and *Invalid_y*.

3. *requested_security_strength* = the security strength associated with the (L, N) pair; see SP 800-57.

4. Obtain a string of $N+64$ *returned_bits* from an **RBG** with a security strength of *requested_security_strength* or more. If an **ERROR** indication is returned, then return an **ERROR** indication, *Invalid_x*, and *Invalid_y*.

5. Convert *returned_bits* to the (non-negative) integer c (see Appendix C.2.1).

6. $x = (c \bmod (q–1)) + 1$.   Comment: $0 \le c \bmod (q–1) \le q–2$ and implies that $1 \le x \le q–1$.

7. $y = g^x \bmod p$.

8. Return **SUCCESS**, x, and y.

B.1.2 Key Pair Generation by Testing Candidates

In this method, a random number is obtained and tested to determine that it will produce a value of x in the correct range. If x is out-of-range, another random number is obtained (i.e., the process is iterated until an acceptable value of x is obtained).

The following process or its equivalent may be used to generate an FFC key pair.

Input:

(p, q, g) The subset of the domain parameters that are used for this process. p, q and g **shall** either be provided as integers during input, or **shall** be converted to integers prior to use.

Output:

1. *status*   The status returned from the key pair generation process. The status will indicate **SUCCESS** or an **ERROR**.

2. *(x, y)*   The generated private and public keys. If an error is encountered during the generation process, invalid values for *x* and *y* **should** be returned, as represented by *Invalid_x* and *Invalid_y* in the following specification. *x* and *y* are returned as integers. The generated private key *x* is in the range [1, *q*–1], and the public key is in the range [1, *p*–1].

Process:

1. $N = \mathbf{len}(q)$; $L = \mathbf{len}(p)$.

> Comment: Check that the (*L*, *N*) pair is specified in Section 4.2.

2. If the (*L*, *N*) pair is invalid, then return an **ERROR** indication, *Invalid_x*, and *Invalid_y*.

3. *requested_security_strength* = the security strength associated with the (*L*, *N*) pair; see SP 800-57.

4. Obtain a string of *N returned_bits* from an **RBG** with a security strength of *requested_security_strength* or more. If an **ERROR** indication is returned, then return an **ERROR** indication, *Invalid_x*, and *Invalid_y*.

5. Convert *returned_bits* to the (non-negative) integer *c* (see Appendix C.2.1).

6. If ($c > q$–2), then go to step 4.

7. $x = c + 1$.

8. $y = g^x \bmod p$.

9. Return **SUCCESS**, *x*, and *y*.

B.2   FFC Per-Message Secret Number Generation

DSA requires the generation of a new random number *k* for each message to be signed. Two methods are provided for the generation of *k*; one of these two methods or another approved method **shall** be used.

The valid values of *N* are provided in Section 4.2. Let **inverse**(*k*, *q*) be a function that computes the inverse of a (non-negative) integer *k* with respect to multiplication modulo the prime number *q*. A technique for computing the inverse is provided in Appendix C.1.

B.2.1 Per-Message Secret Number Generation Using Extra Random Bits

In this method, 64 more bits are requested from the RBG than are needed for k so that bias produced by the mod function in step 6 is not readily apparent.

The following process or its equivalent may be used to generate a per-message secret number.

Input:

    (p, q, g) DSA domain parameters that are generated as specified in Section 4.3.1.

Output:

1. *status*   The status returned from the secret number generation process. The status will indicate **SUCCESS** or an **ERROR**.

2. (k, k^{-1}) The per-message secret number k and its mod q inverse, k^{-1}. If an error is encountered during the generation process, invalid values for k and k^{-1} **should** be returned, as represented by *Invalid_k* and *Invalid_k_inverse* in the following specification. k and k^{-1} are in the range [1, q–1].

Process:

1. $N = \textbf{len}(q)$; $L = \textbf{len}(p)$.

> Comment: Check that the (L, N) pair is specified in Section 4.2.

2. If the (L, N) pair is invalid, then return an **ERROR** indication, *Invalid_k*, and *Invalid_k_inverse*.

3. *requested_security_strength* = the security strength associated with the (L, N) pair; see SP 800-57.

4. Obtain a string of $N+64$ *returned_bits* from an **RBG** with a security strength of *requested_security_strength* or more. If an **ERROR** indication is returned, then return an **ERROR** indication, *Invalid_k*, and *Invalid_k_inverse*.

5. Convert *returned_bits* to the (non-negative) integer c (see Appendix C.2.1).

6. $k = (c \bmod (q–1)) + 1$.

7. $(status, k^{-1}) = \textbf{inverse}\ (k, q)$.

8. Return *status*, k, and k^{-1}.

B.2.2 Per-Message Secret Number Generation by Testing Candidates

In this method, a random number is obtained and tested to determine that it will produce a value of k in the correct range. If k is out-of-range, another random number is obtained (i.e., the process is iterated until an acceptable value of k is obtained.

The following process or its equivalent may be used to generate a per-message secret number.

Input:

(p, q, g) — DSA domain parameters that are generated as specified in Section 4.3.1.

Output:

1. *status* — The status returned from the secret number generation process. The status will indicate **SUCCESS** or an **ERROR**.

2. (k, k^{-1}) — The per-message secret number k and its inverse, k^{-1}. If an error is encountered during the generation process, invalid values for k and k^{-1} **should** be returned, as represented by *Invalid_k* and *Invalid_k_inverse* in the following specification. k and k^{-1} are in the range $[1, q-1]$.

Process:

1. $N = \mathbf{len}(q); L = \mathbf{len}(p)$.

   > Comment: Check that the (L, N) pair is specified in Section 4.2).

2. If the (L, N) pair is invalid, then return an **ERROR** indication, *Invalid_k*, and *Invalid_k_inverse*.

3. *requested_security_strength* = the security strength associated with the (L, N) pair; see SP 800-57.

4. Obtain a string of N *returned_bits* from an **RBG** with a security strength of *requested_security_strength* or more. If an **ERROR** indication is returned, then return an **ERROR** indication, *Invalid_k*, and *Invalid_k_inverse*.

5. Convert *returned_bits* to the (non-negative) integer c (see Appendix C.2.1).

6. If $(c > q-2)$, then go to step 4.

7. $k = c + 1$.

8. $(status, k^{-1}) = \mathbf{inverse}(k, q)$.

9. Return *status*, k, and k^{-1}.

B.3   IFC Key Pair Generation

B.3.1 Criteria for IFC Key Pairs

Key pairs for IFC consist of a public key (n, e), and a private key (n, d), where n is the modulus and is the product of two prime numbers p and q. The security of IFC depends on the quality and secrecy of these primes and the private exponent d. The primes p and q **shall** be generated using one of the following methods:

A. Both p and q are randomly generated prime numbers (Random Primes), where p and q **shall** both be either :

   1. Provable primes (see Appendix B.3.2), or

   2. Probable primes (see Appendix B.3.3).

   Using methods 1 and 2, p and q with lengths of 1024 or 1536 bits may be generated; p and q with lengths of 512 bits **shall not** be generated using these methods. Instead, p and q with lengths of 512 bits **shall** be generated using the conditions based on auxiliary primes (see Appendices B.3.4, B.3.5, or B.3.6).

B. Both p and q are randomly generated prime numbers that satisfy the following additional conditions (Primes with Conditions):

   - $(p-1)$ has a prime factor p_1

   - $(p+1)$ has a prime factor p_2

   - $(q-1)$ has a prime factor q_1

   - $(q+1)$ has a prime factor q_2

   where p_1, p_2, q_1 and q_2 are called auxiliary primes of p and q.

   Using this method, one of the following cases **shall** apply:

   1. The primes p_1, p_2, q_1, q_2, p and q **shall** all be provable primes (see Appendix B.3.4),

   2. The primes p_1, p_2, q_1 and q_2 **shall** be provable primes, and the primes p and q **shall** be probable primes (see Appendix B.3.5), or

   3 The primes p_1, p_2, q_1, q_2, p and q **shall** all be probable primes (see Appendix B.3.6).

The minimum lengths for each of the auxiliary primes p_1, p_2, q_1 and q_2 are dependent on *nlen*, where *nlen* is the length of the modulus n in bits. Note that *nlen* is also called the key size. The lengths of the auxiliary primes may be fixed or randomly chosen, subject to the restrictions in Table B.1. The maximum length is determined by *nlen* (the sum of the length of each auxiliary prime pair) and whether the primes p and q are probable primes or provable primes (e.g., for the auxiliary prime pair p_1 and p_2, **len**(p_1) + **len**(p_2) **shall** be less than a value determined by *nlen*, whether p_1 and p_2 are generated to be probable or provable primes)[3].

---

[3] For the probable primes p and q: **len**(p_1) + **len**(p_2) < **len**(p) − $\log_2$(**len**(p)) − 6; similarly for **len**(q_1) + **len**(q_2) and **len**(q). For the provable primes p and q: **len**(p_1) + **len**(p_2) < **len**(p)/2 − $\log_2$(**len**(p)) − 7; similarly for **len**(q_1) + **len**(q_2)

Table B.1. Minimum and maximum lengths of p_1, p_2, q_1 and q_2

nlen	Min. length of auxiliary primes p_1, p_2, q_1 and q_2	Max. length of len(p_1) + len(p_2) and len(q_1) + len(q_2)	
		p, q Probable primes	p, q Provable primes
1024	> 100 bits	< 496 bits	< 239 bits
2048	> 140 bits	< 1007 bits	< 494 bits
3072	> 170 bits	< 1518 bits	< 750 bits

For different values of *nlen* (i.e., different key sizes), the methods allowed for the generation of p and q are specified in Table B.2.

Table B.2. Allowable Prime Generation Methods

nlen	Random Primes	Primes with Conditions
1024	No	Yes
2048	Yes	Yes
3072	Yes	Yes

In addition, all IFC keys **shall** meet the following criteria in order to conform to FIPS 186-4:

1. The public exponent *e* **shall** be selected with the following constraints:

   (a) The public verification exponent *e* **shall** be selected prior to generating the primes *p* and *q*, and the private signature exponent *d*.

   (b) The exponent *e* **shall** be an odd positive integer such that:

   $$2^{16} < e < 2^{256}.$$

   Note that the value of *e* may be any value that meets constraint 1(b), i.e., *e* may be either a fixed value or a random value.

2. The primes *p* and *q* **shall** be selected with the following constraints:

   (a) (*p*–1) and (*q*–1) **shall** be relatively prime to the public exponent *e*.

   (b) The private prime factor *p* **shall** be selected randomly and **shall** satisfy

---

and **len**(*q*). In each case, **len**(*p*) = **len**(*q*) = *nlen*/2.

$(\sqrt{2})(2^{(nlen/2)-1}) \le p \le (2^{nlen/2}-1)$, where *nlen* is the appropriate length for the desired *security_strength*.

    (c) The private prime factor *q* **shall** be selected randomly and **shall** satisfy $(\sqrt{2})(2^{(nlen/2)-1}) \le q \le (2^{nlen/2}-1)$, where *nlen* is the appropriate length for the desired *security_strength*.

    (d) $|p-q| > 2^{(nlen/2)-100}$.

3. The private signature exponent *d* **shall** be selected with the following constraints after the generation of *p* and *q*:

    (a) The exponent *d* **shall** be a positive integer value such that $2^{nlen/2} < d < \text{LCM}(p-1, q-1)$, and

    (b) $d = e^{-1} \bmod (\text{LCM}(p-1, q-1))$.

        That is, the inequality in (a) holds, and $1 \equiv (ed) \pmod{\text{LCM}(p-1, q-1)}$.

In the extremely rare event that $d \le 2^{nlen/2}$, then new values for *p*, *q* and *d* **shall** be determined. A different value of *e* may be used, although this is not required.

Any hash function used during the generation of the key pair **shall** be approved (i.e., specified in FIPS 180).

B.3.2 Generation of Random Primes that are Provably Prime

An approved method that satisfies the constraints of Appendix B.3.1 **shall** be used for the generation of IFC random primes *p* and *q* that are provably prime (see case A.1). One such method is provided in Appendix B.3.2.1 and B.3.2.2. For this method, a random seed is initially required (see Appendix B.3.2.1); the length of the seed is equal to twice the security strength associated with the modulus *n*. After the seed is obtained, the primes can be generated (see Appendix B.3.2.2).

B.3.2.1 Get the Seed

The following process or its equivalent **shall** be used to generate the seed for this method.

Input:

    nlen   The intended bit length of the modulus *n*.

Output:

    status   The status to be returned, where *status* is either **SUCCESS** or **FAILURE**.

    seed   The seed. If *status* = **FAILURE**, a value of zero is returned as the *seed*.

Process:

1. If *nlen* is not valid (see Section 5.1), then Return (**FAILURE**, 0).

2. Let *security_strength* be the security strength associated with *nlen*, as specified in SP 800-57, Part 1.

3. Obtain a string *seed* of (2 * *security_strength*) bits from an **RBG** that supports the *security_strength*.

4. Return (**SUCCESS**, *seed*).

B.3.2.2 Construction of the Provable Primes *p* and *q*

The following process or its equivalent **shall** be used to construct the random primes *p* and *q* (to be used as factors of the RSA modulus *n*) that are provably prime:

Input:

nlen	The intended bit length of the modulus *n*.
e	The public verification exponent.
seed	The seed obtained using the method in Appendix B.3.2.1.

Output:

status	The status of the generation process, where *status* is either **SUCCESS** or **FAILURE**. When **FAILURE** is returned, zero values **shall** be returned as the other parameters.
p and *q*	The private prime factors of *n*.

Process:

1. If *nlen* is neither 2048 nor 3072, then return (**FAILURE**, 0, 0).

2. If $((e \leq 2^{16})$ OR $(e \geq 2^{256})$ OR (e is not odd)), then return (**FAILURE**, 0, 0).

3. Set the value of *security_strength* in accordance with the value of *nlen*, as specified in SP 800-57, Part 1.

4. If (**len**(*seed*) ≠ 2 * *security_strength*), then return (**FAILURE**, 0, 0).

5. *working_seed* = *seed*.

6. Generate *p*:

   6.1 Using *L* = *nlen*/2, $N_1 = 1$, $N_2 = 1$, *first_seed* = *working_seed* and *e*, use the provable prime construction method in Appendix C.10 to obtain *p* and *pseed*. If **FAILURE** is returned, then return (**FAILURE**, 0, 0).

   6.2 *working_seed* = *pseed*.

54

7. Generate q:

    7.1   Using $L = nlen/2$, $N_1 = 1$, $N_2 = 1$, *first_seed* = *working_seed* and e, use the provable prime construction method in Appendix C.10 to obtain q and *qseed*. If **FAILURE** is returned, then return (**FAILURE**, 0, 0).

    7.2   *working_seed* = *qseed*.

8. If ($|p - q| \leq 2^{nlen/2 - 100}$), then go to step 7.

9. Zeroize the internally generated seeds:

    9.1 *pseed* = 0;

    9.2 *qseed* = 0;

    9.3 *working_seed* = 0.

10. Return (**SUCCESS**, p, q).

B.3.3 Generation of Random Primes that are Probably Prime

An approved method that satisfies the constraints of Appendix B.3.1 **shall** be used for the generation of IFC random primes p and q that are probably prime (see case A.2).

The following process or its equivalent **shall** be used to construct the random probable primes p and q (to be used as factors of the RSA modulus n):

Input:

    nlen   The intended bit length of the modulus n.

    e The public verification exponent.

Output:

    status   The status of the generation process, where *status* is either **SUCCESS** or **FAILURE**.

    p and q The private prime factors of n. When **FAILURE** is returned, zero values **shall** be returned as p and q.

Process:

1. If *nlen* is neither 2048 nor 3072, return (**FAILURE**, 0, 0).

2. If $((e \leq 2^{16})$ OR $(e \geq 2^{256})$ OR (e is not odd)), then return (**FAILURE**, 0, 0).

3. Set the value of *security_strength* in accordance with the value of *nlen*, as specified in SP 800-57, Part 1.

4. Generate p:

    4.1 $i = 0$.

4.2   Obtain a string p of ($nlen/2$) bits from an **RBG** that supports the *security_strength*.

4.3   If (p is not odd), then $p = p + 1$.

4.4   If $((p < (\sqrt{2})(2^{(nlen/2)-1}))$, then go to step 4.2.

4.5   If (**GCD**($p-1$, e) = 1), then

        4.5.1 Test p for primality as specified in Appendix C.3, using an appropriate value from Table C-2 or C-3 in Appendix C.3 as the number of iterations.

        4.5.2 If p is **PROBABLY PRIME**, then go to step 5.

4.6 $i = i + 1$.

4.7   If ($i \geq 5(nlen/2)$), then return (**FAILURE**, 0, 0)

     Else go to step 4.2.

5.   Generate q:

   5.1 $i = 0$.

   5.2   Obtain a string q of ($nlen/2$) bits from an **RBG** that supports the *security_strength*

   5.3   If (q is not odd), then $q = q + 1$.

   5.4   If ($|p - q| \leq 2^{nlen/2 - 100}$), then go to step 5.2.

   5.5   If $((q < (\sqrt{2})(2^{(nlen/2)-1}))$, then go to step 5.2.

   5.6   If (**GCD**($q-1$, e) = 1) then

        5.6.1 Test q for primality as specified in Appendix C.3, using an appropriate value from Table C-2 or C-3 in Appendix C.3 as the number of iterations.

        5.6.2 If q is **PROBABLY PRIME**, then return (**SUCCESS**, p, q).

   5.7 $i = i + 1$.

   5.8   If ($i \geq 5(nlen/2)$), then return (**FAILURE**, 0, 0)

     Else go to step 5.2.

B.3.4 Generation of Provable Primes with Conditions Based on Auxiliary Provable Primes

This section specifies an approved method for the generation of the IFC primes p and q with the additional conditions specified in Appendix B.3.1, case B.1, where p, p_1, p_2, q, q_1 and q_2 are all provable primes. For this method, a random seed is initially required (see Appendix B.3.2.1); the length of the seed is equal to twice the security strength associated with the modulus n. After the first seed is obtained, the primes can be generated.

Let $bitlen_1$, $bitlen_2$, $bitlen_3$, and $bitlen_4$ be the bit lengths for p_1, p_2, q_1 and q_2, respectively, in accordance with Table B.1. The following process or its equivalent **shall** be used to generate the provable primes:

Input:

nlen	The intended bit length of the modulus n.
e	The public verification exponent.
seed	The seed obtained using the method in Appendix B.3.2.1.

Output:

status	The status of the generation process, where *status* is either **SUCCESS** or **FAILURE**. If **FAILURE** is returned then zeros **shall** be returned as the values for p and q.
p and *q*	The private prime factors of n.

Process:

1. If *nlen* is neither 1024, 2048, nor 3072, then return (**FAILURE**, 0, 0).

2. If $((e \leq 2^{16})$ OR $(e \geq 2^{256})$ OR (e is not odd)), then return (**FAILURE**, 0, 0).

3. Set the value of *security_strength* in accordance with the value of *nlen*, as specified in SP 800-57, Part 1.

4. If (**len**(*seed*) $\neq$ 2 * *security_strength*), then return (**FAILURE**, 0, 0).

5. *working_seed* = *seed*.

6. Generate p:

   6.1 Using $L = nlen/2$, $N_1 = bitlen_1$, $N_2 = bitlen_2$, *firstseed* = *working_seed* and e, use the provable prime construction method in Appendix C.10 to obtain p, p_1, p_2 and *pseed*. If **FAILURE** is returned, return (**FAILURE**, 0, 0).

   6.2 *working_seed* = *pseed*.

7. Generate q:

   7.1 Using $L = nlen/2$, $N_1 = bitlen_3$, $N_2 = bitlen_4$ and *firstseed* = *working_seed* and e, use the provable prime construction method in Appendix C.10 to obtain q, q_1, q_2 and *qseed*. If **FAILURE** is returned, return (**FAILURE**, 0, 0).

   7.2 *working_seed* = *qseed*.

8. If ($|p - q| \leq 2^{nlen/2 - 100}$), then go to step 7.

9. Zeroize the internally generated seeds:

   9.1 *pseed* = 0.

9.2 *qseed* = 0.

9.3 *working_seed* = 0.

10. Return (**SUCCESS**, *p*, *q*).

B.3.5 Generation of Probable Primes with Conditions Based on Auxiliary Provable Primes

This section specifies an approved method for the generation of the IFC primes *p* and *q* with the additional conditions specified in Appendix B.3.1, case B.2, where p_1, p_2, q_1 and q_2 are provably prime, and *p* and *q* are probably prime. For this method, a random seed is initially required (see Appendix B.3.2.1); the length of the seed is equal to twice the security strength associated with the modulus *n*. After the first seed is obtained, the primes can be generated.

Let $bitlen_1$, $bitlen_2$, $bitlen_3$, and $bitlen_4$ be the bit lengths for p_1, p_2, q_1 and q_2, respectively in accordance with Table B.1. The following process or its equivalent **shall** be used to construct *p* and *q*.

Input:

nlen	The intended bit length of the modulus *n*.
e	The public verification exponent.
seed	The seed obtained using the method in Appendix B.3.2.1.

Output:

status	The status of the generation process, where *status* is either **SUCCESS** or **FAILURE**. If **FAILURE** is returned then zeros **shall** be returned as the values for *p* and *q*.
p and *q*	The private prime factors of *n*.

Process:

1. If *nlen* is neither 1024, 2048, nor 3072, then return (**FAILURE**, 0, 0).

2. If $((e \leq 2^{16})$ OR $(e \geq 2^{256})$ OR (*e* is not odd)), then return (**FAILURE**, 0, 0).

3. Set the value of *security_strength* in accordance with the value of *nlen*, as specified in SP 800-57, Part 1.

4. If (**len**(*seed*) ≠ 2 * *security_strength*), then return (**FAILURE**, 0, 0).

> Comment: Generate four primes p_1, p_2, q_1 and q_2 that are provably prime.

5. Generate p:

  5.1 Using $bitlen_1$ as the length, and *seed* as the *input_seed*, use the random prime generation routine in Appendix C.6 to obtain p_1 and *prime_seed*. If **FAILURE** is returned, the return (**FAILURE**, 0, 0).

  5.2 Using $bitlen_2$ as the length, and *prime_seed* as the *input_seed*, use the random prime generation routine in Appendix C.6 to obtain p_2 and a new value for *prime_seed*. If **FAILURE** is returned, the return (**FAILURE**, 0, 0).

  5.3 Generate a prime p using the routine in Appendix C.9 with inputs of p_1, p_2, *nlen, e* and *security_strength*, also obtaining X_p. If **FAILURE** is returned, return (**FAILURE**, 0, 0).

6. Generate q:

  6.1. Using $bitlen_3$ as the length, and *prime_seed* as the *input_seed*, use the random prime generation routine in Appendix C.6 to obtain q_1 and a new value for *prime_seed*. If **FAILURE** is returned, the return (**FAILURE**, 0, 0).

  6.2 Using $bitlen_4$ as the length, and *prime_seed* as the *input_seed*, use the random prime generation routine in Appendix C.6 to obtain q_2 and a new value for *prime_seed*. If **FAILURE** is returned, the return (**FAILURE**, 0, 0).

  6.3 Generate a prime q using the routine in Appendix C.9 with inputs of q_1, q_2, *nlen, e* and *security_strength*, also obtaining X_q. If **FAILURE** is returned, return (**FAILURE**, 0, 0).

7. If $((|p - q| \leq 2^{nlen/2 - 100})$ OR $(|X_p - X_q| \leq 2^{nlen/2 - 100}))$, then go to step 6.

8. Zeroize the internally generated that are not returned:

  8.1 $X_p = 0$.

  8.2 $X_q = 0$.

  8.3 *prime_seed* $= 0$.

  8.4 $p_1 = 0$.

  8.5 $p_2 = 0$.

  8.6 $q_1 = 0$.

  8.7 $q_2 = 0$.

9. Return (**SUCCESS**, p, q).

B.3.6 Generation of Probable Primes with Conditions Based on Auxiliary Probable Primes

An approved method that satisfies the constraints of Appendix B.3.1 **shall** be used for the generation of IFC primes p and q that are probably prime and meet the additional constraints of Appendix B.3.1 (see case B.3). For this case, the prime factors p_1, p_2, q_1 and q_2 are also probably prime.

Four random numbers X_{p1}, X_{p2}, X_{q1} and X_{q2} are generated, from which the prime factors p_1, p_2, q_1 and q_2 are determined. p_1 and p_2, and an additional random number X_p are then used to determine p, and q_1 and q_2 and a random number X_q are used to obtain q. Let $bitlen_1$, $bitlen_2$, $bitlen_3$, and $bitlen_4$ be the bit lengths for p_1, p_2, q_1 and q_2, respectively chosen in accordance with Table B.1.

The following process or its equivalent **shall** be used to generate p and q:

Input:

nlen	The intended bit length of the modulus n.
e	The public verification exponent.

Output:

status	The status of the generation process, where *status* is either **SUCCESS** or **FAILURE**. If **FAILURE** is returned then zeros **shall** be returned as the values for p and q.
p and q	The private prime factors of n.

Process:

1. If *nlen* is neither 1024, 2048, nor 3072, then return (**FAILURE**, 0, 0).

2. If $((e \le 2^{16})$ OR $(e \ge 2^{256})$ OR (e is not odd)), then return (**FAILURE**, 0, 0).

3. Set the value of *security_strength* in accordance with the value of *nlen*, as specified in SP 800-57, Part 1.

4. Generate p:

    4.1 Generate an odd integer X_{p1} of length $bitlen_1$ bits, and a second odd integer X_{p2} of length $bitlen_2$ bits, using an approved random number generator that supports the *security_strength*.

    4.2 Sequentially search successive odd integers, starting at X_{p1} until the first probable prime p_1 is found. Candidate integers **shall** be tested for primality as specified in Appendix C.3. Repeat the process to find p_2, starting at X_{p2}. The probable primes p_1 and p_2 **shall** be the first integers that pass the primality test.

4.3 Generate a prime p using the routine in Appendix C.9 with inputs of p_1, p_2, $nlen$, e and $security\_ strength$, also obtaining X_p. If **FAILURE** is returned, return (**FAILURE**, 0, 0).

5. Generate q:

   5.1 Generate an odd integer X_{q1} of length $bitlen_3$ bits, and a second odd integer X_{q2} of length $bitlen_4$ bits, using an approved random number generator that supports the $security\_strength$.

   5.2 Sequentially search successive odd integers, starting at X_{q1} until the first probable prime q_1 is found. Candidate integers **shall** be tested for primality as specified in Appendix C.3. Repeat the process to find q_2, starting at X_{q2}. The probable primes q_1 and q_2 **shall** be the first integers that pass the primality test.

   5.3 Generate a prime q using the routine in Appendix C.9 with inputs of q_1, q_2, $nlen$, e and $security\_ strength$, also obtaining X_q. If **FAILURE** is returned, return (**FAILURE**, 0, 0).

6. If $((|X_p - X_q| \leq 2^{nlen/2 - 100})$ OR $(|p - q| \leq 2^{nlen/2 - 100}))$, then go to step 5.

7. Zeroize the internally generated values that are not returned:

   7.1 $X_p = 0$.

   7.2 $X_q = 0$.

   7.3 $X_{p1} = 0$.

   7.4 $X_{p2} = 0$.

   7.5 $X_{q1} = 0$.

   7.6 $X_{q2} = 0$.

   7.7 $p_1 = 0$.

   7.8 $p_2 = 0$.

   7.9 $q_1 = 0$.

   7.10 $q_2 = 0$.

8. Return (**SUCCESS**, p, q).

B.4 ECC Key Pair Generation

An ECC key pair d and Q is generated for a set of domain parameters (q, FR, a, b {, $domain\_parameter\_seed$}, G, n, h). Two methods are provided for the generation of the ECC private key d and public key Q; one of these two methods **shall** be used to generate d and Q.

Prior to generating ECDSA key pairs, assurance of the validity of the domain parameters (q, FR, a, b {, *domain_parameter_seed*}, G, n, h) **shall** have been obtained as specified in Section 3.1.

For ECDSA, the valid bit-lengths of n are provided in Section 6.1.1. See ANS X9.62 for definitions of the elliptic curve math and the conversion routines.

B.4.1 Key Pair Generation Using Extra Random Bits

In this method, 64 more bits are requested from the RBG than are needed for d so that bias produced by the mod function in step 6 is negligible.

The following process or its equivalent may be used to generate an ECC key pair.

Input:

1. (q, FR, a, b {, *domain_parameter_seed*}, G, n, h)

    The domain parameters that are used for this process. n is a prime number, and G is a point on the elliptic curve.

Output:

1. *status*   The status returned from the key pair generation procedure. The status will indicate **SUCCESS** or an **ERROR**.

2. (d, Q)   The generated private and public keys. If an error is encountered during the generation process, invalid values for d and Q **should** be returned, as represented by *Invalid_d* and *Invalid_Q* in the following specification. d is an integer, and Q is an elliptic curve point. The generated private key d is in the range [1, $n–1$].

Process:

1. $N = \textbf{len}(n)$.

                    Comment: Check that N is included in Table 1 of Section 6.1.1.

2. If N is invalid, then return an **ERROR** indication, *Invalid_d*, and *Invalid_Q*.

3. *requested_security_strength* = the security strength associated with N; see SP 800-57, Part 1.

4. Obtain a string of $N+64$ *returned_bits* from an **RBG** with a security strength of *requested_security_strength* or more. If an **ERROR** indication is returned, then return an **ERROR** indication, *Invalid_d*, and *Invalid_Q*.

5. Convert *returned_bits* to the (non-negative) integer c (see Appendix C.2.1).

6. $d = (c \bmod (n–1)) + 1$.

7. $Q = dG$.

8. Return **SUCCESS**, d, and Q.

B.4.2 Key Pair Generation by Testing Candidates

In this method, a random number is obtained and tested to determine that it will produce a value of d in the correct range. If d is out-of-range, another random number is obtained (i.e., the process is iterated until an acceptable value of d is obtained.

The following process or its equivalent may be used to generate an ECC key pair.

Input:

1. $(q, FR, a, b \{, domain\_parameter\_seed\}, G, n, h)$

> The domain parameters that are used for this process. n is a prime number, and G is a point on the elliptic curve.

Output:

1. *status* The status returned from the key pair generation procedure. The status will indicate **SUCCESS** or an **ERROR**.

2. (d, Q) The generated private and public keys. If an error is encountered during the generation process, invalid values for d and Q **should** be returned, as represented by *Invalid_d* and *Invalid_Q* in the following specification. d is an integer, and Q is an elliptic curve point. The generated private key d is in the range $[1, n-1]$.

Process:

1. $N = \mathbf{len}(n)$.

> Comment: Check that N is included in Table 1 of Section 6.1.1.

2. If N is invalid, then return an **ERROR** indication, *Invalid_d*, and *Invalid_Q*.

3. *requested_security_strength* = the security strength associated with N; see SP 800-57, Part 1.

4. Obtain a string of N *returned_bits* from an **RBG** with a security strength of *requested_security_strength* or more. If an **ERROR** indication is returned, then return an **ERROR** indication, *Invalid_d*, and *Invalid_Q*.

5. Convert *returned_bits* to the (non-negative) integer c (see Appendix C.2.1).

6. If $(c > n-2)$, then go to step 4.

7. $d = c + 1$.

8. $Q = dG$.

9. Return **SUCCESS**, d, and Q.

B.5   ECC Per-Message Secret Number Generation

ECDSA requires the generation of a new random number k for each message to be signed. Two methods are provided for the generation of k; one of these two methods or another approved method **shall** be used.

The valid values of n are provided in Section 6.1.1. See ANS X9.62 for definitions of the elliptic curve math and the conversion routines.

Let **inverse**(k, n) be a function that computes the inverse of a (non-negative) integer k with respect to multiplication modulo the prime number n. A technique for computing the inverse is provided in Appendix C.1.

B.5.1 Per-Message Secret Number Generation Using Extra Random Bits

In this method, 64 more bits are requested from the RBG than are needed for k so that bias produced by the mod function in step 6 is not readily apparent.

The following process or its equivalent may be used to generate a per-message secret number.

> **Input:**
>
> 1. $(q, FR, a, b \{, domain\_parameter\_seed\}, G, n, h)$
>
>> The domain parameters that are used for this process. n is a prime number, and G is a point on the elliptic curve.
>
> **Output:**
>
> 1. *status*   The status returned from the key pair generation procedure. The status will indicate **SUCCESS** or an **ERROR**.
>
> 2. (k, k^{-1}) The generated secret number k and its inverse k^{-1}. If an error is encountered during the generation process, invalid values for k and k^{-1} **should** be returned, as represented by *Invalid_k* and *Invalid_k_inverse* in the following specification. k and k^{-1} are integers in the range $[1, n{-}1]$.
>
> **Process:**
>
> 1. $N = \textbf{len}(n)$.
>
>> Comment: Check that N is included in Table 1 of Section 6.1.1.
>
> 2. If N is invalid, then return an **ERROR** indication, *Invalid_k*, and *Invalid_k_inverse*.

3. *requested_security_strength* = the security strength associated with *N*; see SP 800-57, Part 1.

4. Obtain a string of *N*+64 *returned_bits* from an **RBG** with a security strength of *requested_security_strength* or more. If an **ERROR** indication is returned, then return an **ERROR** indication, *Invalid_k*, and *Invalid_k_inverse*.

5. Convert *returned_bits* to the non-negative integer *c* (see Appendix C.2.1).

6. $k = (c \bmod (n-1)) + 1$.

7. $(status,\ k^{-1})$ = **inverse**(*k*, *n*).

8. Return *status*, *k*, and k^{-1}.

B.5.2 Per-Message Secret Number Generation by Testing Candidates

In this method, a random number is obtained and tested to determine that it will produce a value of *k* in the correct range. If *k* is out-of-range, another random number is obtained (i.e., the process is iterated until an acceptable value of *k* is obtained.

The following process or its equivalent may b used to generate a per-message secret number.

Input:

1. (*q*, *FR*, *a*, *b* {, *domain_parameter_seed*}, *G*, *n*, *h*)

    The domain parameters that are used for this process. *n* is a prime number, and *G* is a point on the elliptic curve.

Output:

1. *status*   The status returned from the key pair generation procedure. The status will indicate **SUCCESS** or an **ERROR**.

2. (k, k^{-1}) The generated secret number *k* and its inverse k^{-1}. If an error is encountered during the generation process, invalid values for *k* and k^{-1} **should** be returned, as represented by *Invalid_k* and *Invalid_k_inverse* in the following specification. *k* and k^{-1} are integers in the range [1, *n*–1].

Process:

1. $N = \textbf{len}(n)$.

    Comment: Check that *N* is included in Table 1 of Section 6.1.1.

2. If *N* is not included in Table 1, then return an **ERROR** indication, *Invalid_k*, and *Invalid_k_inverse*.

3. *requested_security_strength* = the security strength associated with *N*; see SP 800-57, Part 1.

4. Obtain a string of *N returned_bits* from an **RBG** with a security strength of *requested_security_strength* or more. If an **ERROR** indication is returned, then return an **ERROR** indication, *Invalid_k*, and *Invalid_k_inverse*.

5. Convert *returned_bits* to the (non-negative) integer *c* (see Appendix C.2.1).

6. If ($c > n-2$), then go to step 4.

7. $k = c + 1$.

8. ($status$, k^{-1}) = **inverse**(k, n).

9. Return *status*, k, and k^{-1}.

Appendix C: Generation of Other Quantities

This appendix contains routines for supplementary processes required for the implementation of this Standard. Appendix C.1 is needed to produce the inverse of the per-message secret k (see Section 4.5, and Appendices B.2.1, B.2.2, B.5.1 and B.5.2) and the inverse of the signature portion s that is used during signature verification (see Section 4.7). The routines in Appendix C.2 are required to convert between bit strings and integers where required in implementing this Standard. Appendix C.3 contains probabilistic primality tests to be used during the generation of DSA domain parameters and RSA key pairs. Appendices C.4 and C.5 contain algorithms required during the Lucas probabilistic primality test of Appendix C.3.3 to check for a perfect square and to compute the Jacobi symbol. Appendix C.6 contains the Shawe-Taylor algorithm for the construction of primes. Appendix C.7 provides a process to perform trial division, as required by the random prime generation routine in Appendix C.6. The sieve procedure in Appendix C.8 is needed by the trial division routine in Appendix C.7. The trial division process in Appendix C.7 and the sieve procedure in Appendix C.8 have been extracted from ANS X9.80, *Prime Number Generation, Primality Testing, and Primality Certificates*. Appendix C.9 is required during the generation of RSA key pairs. Appendix C.10 provides a method for constructing provable primes for RSA (see Appendix B.3.2.2 and B.3.4).

C.1 Computation of the Inverse Value

This algorithm or an algorithm that produces an equivalent result **shall** be used to compute the multiplicative inverse z^{-1} mod a, where $0 < z < a$, $0 < z^{-1} < a$, and a is a prime number. In this Standard, z is either k or s, and a is either q or n.

Input:

    1. z The value to be inverted mod a (i.e., either k or s).

    2. a The domain parameter and (prime) modulus (i.e., either q or n).

Output:

    1. *status*   The status returned from this function, where the *status* is either **SUCCESS** or **ERROR**.

    2. z^{-1} The multiplicative inverse of z mod a, if it exists.

Process:

    1. Verify that a and z are positive integers such that $z < a$; if not, return an **ERROR** indication.

    2. Set $i = a$, $j = z$, $y_2 = 0$, and $y_1 = 1$.

    3. *quotient* $= \lfloor i/j \rfloor$.

4. *remainder = i −(j * quotient)*.

5. *y = y₂ −(y₁ * quotient)*.

   5. $y = y_2 - (y_1 * quotient)$.

6. Set $i = j$, $j = remainder$, $y_2 = y_1$, and $y_1 = y$.

7. If (j > 0), then go to step 3.

8. If ($i \neq 1$), then return an **ERROR** indication.

9. Return **SUCCESS** and y_2 mod a.

C.2   Conversion Between Bit Strings and Integers

C.2.1 Conversion of a Bit String to an Integer

An *n*-long sequence of bits { x_1, ..., x_n } is converted to an integer by the rule

$$\{ x_1, \ldots , x_n \} \rightarrow (x_1 * 2^{n-1}) + (x_2 * 2^{n-2}) + \ldots + (n_1 * 2) + x_n .$$

Note that the first bit of a sequence corresponds to the most significant bit of the corresponding integer, and the last bit corresponds to the least significant bit.

Input:

1. $b_1, b_2, \ldots , b_n$ The bit string to be converted.

Output:

1. C The requested integer representation of the bit string.

Process:

1. Let $(b_1, b_2, \ldots , b_n)$ be the bits of b from leftmost to rightmost.

2. $C = \sum_{i=1}^{n} 2^{(n-i)} b_i$

3. Return C.

In this Standard, the binary length of an integer C is defined as the smallest integer n satisfying $C < 2^n$.

C.2.2 Conversion of an Integer to a Bit String

An integer x in the range $0 \leq x < 2^n$ may be converted to an *n*-long sequence of bits by using its binary expansion as shown below:

$$x = (x_1 * 2^{n-1}) + (x_2 * 2^{n-2}) + \ldots + (x_{n-1} * 2) + x_n \rightarrow \{x_1, \ldots, x_n\}$$

Note that the first bit of a sequence corresponds to the most significant bit of the corresponding integer, and the last bit corresponds to the least significant bit.

Input:

    1. C The non-negative integer to be converted.

Output:

    1. $b_1, b_2, \ldots, b_n$ The bit string representation of the integer C.

Process:

1. Let $(b_1, b_2, \ldots, b_n)$ represent the bit string, where $b_i = 0$ or 1, and b_1 is the most significant bit, while b_n is the least significant bit.

2. For any integer n that satisfies $C < 2^n$, the bits b_i **shall** satisfy:

$$C = \sum_{i=1}^{n} 2^{(n-i)} b_i$$

3. Return $b_1, b_2, \ldots, b_n$.

In this Standard, the binary length of the integer C is defined as the smallest integer n that satisfies $C < 2^n$.

C.3   Probabilistic Primality Tests

A probabilistic primality test may be required during the generation and validation of prime numbers. An approved robust probabilistic primality test **shall** be selected and used.

There are several probabilistic algorithms available. The Miller-Rabin probabilistic primality tests described in Appendices C.3.1 and C.3.2 are versions of a procedure due to M.O. Rabin, based in part on ideas of Gary L. Miller; one of these versions **shall** be used as the Miller-Rabin test discussed below. For more information, see [4]. For these tests, let **RBG** be an approved random bit generator.

There are several Lucas probabilistic primality tests available; the version provided in [5] is specified in Appendix C.3.3.

This Standard allows two alternatives for testing primality: either using several iterations of only the Miller-Rabin test, or using the iterated Miller-Rabin test, followed by a single Lucas test. The value of *iterations* (as used in Appendices C.3.1 and C.3.2) depends on the algorithm being used, the security strength, the error probability used, the length (in bits) of the candidate prime and the type of tests to be performed. Tables C.1, C.2 and C.3 list the minimum number of *iterations* of the Miller-Rabin tests that **shall** be performed.

As stated in Appendix F, if the definition of the error probability that led to the values of the number of Miller-Rabin tests for *p* and *q* in Tables C.1, C.2 and C.3 is not conservative enough, the prescribed number of Miller-Rabin tests can be followed by a single Lucas test. Since there are no known non-prime values that pass the two test combination (i.e., the indicated number of rounds of the Miller-Rabin test with randomly selected bases, followed by one round of the Lucas test), the two test combination may provide additional assurance of primality over the use of only the Miller-Rabin test. For DSA, the two-test combination may provide better performance. However, the Lucas test is not required when testing the p_1, p_2, q_1 and q_2 values for primality when generating RSA primes. See Appendix F for further information.

Table C.1. Minimum number of Miller-Rabin iterations for DSA

Parameters	M-R Tests Only	M-R Tests when followed by One Lucas test
p: 1024 bits *q*: 160 bits Error probability = 2^{-80}	For *p* and *q*: 40	For *p*: 3 For *q*: 19
p: 2048 bits *q*: 224 bits Error probability = 2^{-112}	For *p* and *q*: 56	For *p*: 3 For *q*: 24
p: 2048 bits *q*: 256 bits Error probability = 2^{-112}	For *p* and *q*: 56	For *p*: 3 For *q*: 27
p: 3072 bits *q*: 256 bits Error probability = 2^{-128}	For *p* and *q*: 64	For *p*: 2 For *q*: 27

Table C.2. Minimum number of rounds of M-R testing when generating primes for use in RSA Digital Signatures

Parameters	M-R Tests Only
p_1, p_2, q_1 and q_2 > 100 bits p and q: 512 bits Error probability = 2^{-80}	For p_1, p_2, q_1 and q_2: 28 For p and q: 5
p_1, p_2, q_1 and q_2 > 140 bits p and q: 1024 bits Error probability = 2^{-112}	For p_1, p_2, q_1 and q_2: 38 For p and q: 5
p_1, p_2, q_1 and q_2 > 170 bits p and q: 1536 bits Error probability = 2^{-128}	For p_1, p_2, q_1 and, q_2: 41 For p and q: 4

Table C.3. Minimum number of rounds of M-R testing when generating primes for use in RSA Digital Signatures using an error probability of 2^{-100}

Parameters	M-R Tests Only
p_1, p_2, q_1 and q_2 > 100 bits p and q: 512	For p_1, p_2, q_1 and q_2: 38 For p and q: 7
p_1, p_2, q_1 and q_2 > 140 bits p and q: 1024 bits	For p_1, p_2, q_1 and q_2: 32 For p and q: 4
p_1, p_2, q_1 and q_2 > 170 bits p and q: 1536 bits	For p_1, p_2, q_1 and q_2: 27 For p and q: 3

C.3.1 Miller-Rabin Probabilistic Primality Test

Let **RBG** be an approved random bit generator.

Input:

1. w The odd integer to be tested for primality. This will be either p or q, or one of the auxiliary primes p_1, p_2, q_1 or q_2.

2.	*iterations*	The number of iterations of the test to be performed; the value **shall** be consistent with Table C.1, C.2 or C.3.

Output:

1.	*status*	The status returned from the validation procedure, where *status* is either **PROBABLY PRIME** or **COMPOSITE**.

Process:

1. Let a be the largest integer such that 2^a divides $w-1$.

2. $m = (w-1) / 2^a$.

3. $wlen = \textbf{len}(w)$.

4. For $i = 1$ to *iterations* do

    4.1   Obtain a string b of *wlen* bits from an RBG.

                Comment: Ensure that $1 < b < w-1$.

    4.2   If $((b \le 1)$ or $(b \ge w-1))$, then go to step 4.1.

    4.3 $z = b^m \bmod w$.

    4.4   If $((z = 1)$ or $(z = w - 1))$, then go to step 4.7.

    4.5   For $j = 1$ to $a - 1$ do.

        4.5.1 $z = z^2 \bmod w$.

        4.5.2   If $(z = w-1)$, then go to step 4.7.

        4.5.3   If $(z = 1)$, then go to step 4.6.

    4.6 Return **COMPOSITE.**

    4.7 Continue.   Comment: Increment i for the do-loop in step 4.

5. Return **PROBABLY PRIME.**

C.3.2 Enhanced Miller-Rabin Probabilistic Primality Test

This method provides additional information when an error is encountered that may be useful when generating or validating RSA moduli. Let **RBG** be an approved random bit generator.

    Input:

1.	w	The odd integer to be tested for primality. This will be either p or q, or one of the auxiliary primes p_1, p_2, q_1 or q_2.

2.	*iterations*	The number of iterations of the test to be performed; the value **shall** be consistent with Table C.1, C.2 or C.3.

Output:

1.	*status*	The status returned from the validation procedure, where *status* is either **PROBABLY PRIME, PROVABLY COMPOSITE WITH FACTOR** (returned with the factor), and **PROVABLY COMPOSITE AND NOT A POWER OF A PRIME.**

Process:

1. Let a be the largest integer such that 2^a divides $w–1$.

2. $m = (w–1) / 2^{a.}$

3. $wlen = \textbf{len}\ (w)$.

4. For $i = 1$ to *iterations* do

    4.1   Obtain a string b of $wlen$ bits from an RBG.

                        Comment: Ensure that $1 < b < w–1$.

    4.2   If $((b \leq 1)$ or $(b \geq w–1))$, then go to step 4.1.

    4.3 $g = \textbf{GCD}(b, w)$.

    4.4   If $(g > 1)$, then return **PROVABLY COMPOSITE WITH FACTOR** and the value of g.

    4.5 $z = b^m \bmod w$.

    4.6   If $((z = 1)$ or $(z = w - 1))$, then go to step 4.15.

    4.7   For $j = 1$ to $a - 1$ do.

        4.7.1 $x = z$.   Comment: $x \neq 1$ and $x \neq w–1$.

        4.7.2 $z = x^2 \bmod w$.

        4.7.3   If $(z = w–1)$, then go to step 4.15.

        4.7.4   If $(z = 1)$, then go to step 4.12.

    4.8 $x = z$.   Comment: $x = b^{(w–1)/2} \bmod w$ and $x \neq w–1$.

    4.9 $z = x^2 \bmod w$.

    4.10 If $(z = 1)$, then go to step 4.12.

    4.11 $x = z$.   Comment: $x = b^{(w–1)} \bmod w$ and $x \neq 1$.

4.12 $g = \mathbf{GCD}(x{-}1, w)$.

4.13 If $(g > 1)$, then return **PROVABLY COMPOSITE WITH FACTOR** and the value of g.

4.14 Return **PROVABLY COMPOSITE AND NOT A POWER OF A PRIME.**

4.15 Continue.   Comment: Increment i for the do-loop in step 4.

5. Return **PROBABLY PRIME.**

C.3.3 (General) Lucas Probabilistic Primality Test

The following process or its equivalent **shall** be used as the Lucas test.

Input:

  C The candidate odd integer to be tested for primality.

Output:

  status Where *status* is either **PROBABLY PRIME** or **COMPOSITE**.

Process:

1. Test whether C is a perfect square (see Appendix C.4). If so, return (**COMPOSITE**).

2. Find the first D in the sequence $\{5, -7, 9, -11, 13, -15, 17, \ldots\}$ for which the Jacobi symbol $\left(\frac{D}{C}\right) = -1$. See Appendix C.5 for an approved method to compute the Jacobi Symbol. If $\left(\frac{D}{C}\right) = 0$ for any D in the sequence, return (**COMPOSITE**).

3. $K = C{+}1$.

4. Let $K_r K_{r-1} \ldots K_0$ be the binary expansion of K, with $K_r = 1$.

5. Set $U_r = 1$ and $V_r = 1$.

6. For $i = r{-}1$ to 0, do

    6.1 $U_{temp} = U_{i+1} V_{i+1} \bmod C$.

    6.2 $V_{temp} = \dfrac{V_{i+1}{}^2 + DU_{i+1}{}^2}{2} \bmod C.$

    6.3 If $(K_i = 1)$, then   Comment: If $K_i = 1$, then do steps 6.3.1 and 6.3.2; otherwise, do steps 6.3.3 and 6.3.4.

        6.3.1 $U_i = \dfrac{U_{temp} + V_{temp}}{2} \bmod C.$

74

$$6.3.2 \quad V_i = \frac{V_{temp} + DU_{temp}}{2} \bmod C.$$

Else

$$6.3.3 \quad U_i = U_{temp}.$$

$$6.3.4 \quad V_i = V_{temp}.$$

7. If ($U_0 = 0$), then return (**PROBABLY PRIME**). Otherwise, return (**COMPOSITE**).

Steps 6.2, 6.3.1 and 6.3.2 contain expressions of the form $A/2 \bmod C$, where A is an integer, and C is an odd integer. If $A/2$ is not an integer (i.e., A is odd), then $A/2 \bmod C$ may be calculated as $(A+C)/2 \bmod C$. Alternatively, $A/2 \bmod C = A \cdot (C+1)/2 \bmod C$, for any integer A, without regard to A being odd or even.

C.4 Checking for a Perfect Square

The following algorithm may be used to determine whether an n-bit positive integer C is a perfect square:

Input:

> C The integer to be checked.

Output:

> *status*   Where *status* is either **PERFECT SQUARE** or **NOT A PERFECT SQUARE**.

Process:

1. Set n, such that $2^n > C \geq 2^{(n-1)}$.

2. $m = \lceil n/2 \rceil$.

3. $i = 0$.

4. Select X_0, such that $2^m > X_0 \geq 2^{(m-1)}$.

5. Repeat

> 5.1 $i = i + 1$.

> 5.2 $X_i = ((X_{i-1})^2 + C)/(2X_{i-1})$.

Until $(X_i)^2 < 2^m + C$.

6. If $C = \lfloor X_i \rfloor^2$, then

> *status* = **PERFECT SQUARE**.

Else

> *status* = **NOT A PERFECT SQUARE**.

7. **Return** *status*.

Notes:

1. By starting with $X_0 > (1/2)$ **Sqrt**(C), $|X_0 - $**Sqrt**$(C)|$ is guaranteed to be less than X_0. This inequality is maintained in step 5; i.e., $|X_i - $**Sqrt**$(C)| < X_i$ for all i.

2. For $i \geq 1$, $0 \leq X_i - $**Sqrt**$(C) = (X_{i-1} - $**Sqrt**$(C))^2 / (2 X_{i-1}) < X_0/2^i$.

   In particular, $0 \leq X_m - $**Sqrt**$(C) < 1$. If **Sqrt**$(C)$ were an integer, then it would be equal to the floor of X_m.

3. In general, the inequality $X_i - $**Sqrt**$(C) < 1$ will occur for values of i that are much less than m. To detect this, the fact that $2^{(m-1)} \leq $**Sqrt**$(C) < X_i$ for all $i \geq 1$ can be used,

$$X_i - \textbf{Sqrt}(C) = ((X_i)^2 - C)/(X_i + \textbf{Sqrt}(C))$$
$$\leq ((X_i)^2 - C)/(2\,\textbf{Sqrt}(C))$$
$$\leq ((X_i)^2 - C)/(2^m)$$

Thus, the condition $(X_i)^2 < 2^m + C$ implies that $X_i - $**Sqrt**$(C) < 1$.

C.5   Jacobi Symbol Algorithm

This routine computes the Jacobi symbol $\left(\dfrac{a}{n}\right)$.

Jacobi( ):

   Input:

      a Any integer. For this Standard, the initial value is in the sequence $\{5, -7, 9, -11, 13, -15, 17, \ldots\}$, as determined by Appendix C.3.3.

      n Any integer. For this Standard, the initial value is the candidate being tested, as determined by Appendix C.3.3.

   Output:

      result The calculated Jacobi symbol.

   Process:

     1. $a = a \bmod n$.   Comment: a will be in the range $0 \leq a < $ n.

     2. If $a = 1$, or $n = 1$, then return (1).

     3. If $a = 0$, then return (0).

     4. Define e and a_1 such that $a = 2^e a_1$, where a_1 is odd.

5. If e is even, then $s = 1$.

    Else if $((n \equiv 1 \pmod 8))$ or $(n \equiv 7 \pmod 8)))$, then $s = 1$.

    Else if $((n \equiv 3 \pmod 8))$ or $(n \equiv 5 \pmod 8)))$, then $s = -1$.

6. If $((n \equiv 3 \pmod 4))$ and $(a_1 \equiv 3 \pmod 4)))$, then $s = -s$.

7. $n_1 = n \bmod a_1$.

8. Return $(s * \text{Jacobi }(n_1, a_1))$.   Comment: Call this routine recursively.

Example: Compute the Jacobi symbol for $a = 5$ and $n = 3439601197$:

1. n is not 1, and a is not 1, so proceed to Step 2.

2. a is not 0, so proceed to Step 3.

3. $5 = 2^0 * 5$, so $e = 0$, and $a_1 = 5$.

4. e is even, so $s = 1$.

5. a_1 is not congruent to 3 mod 4, so do not change s.

6. $n_1 = 2 = n \bmod 5$.

7. Compute and return $(1 * \text{Jacobi}(2, 5))$. This calls Jacobi recursively. Compute the Jacobi symbol for $a = 2$ and $n = 5$:

    7.1 n is not 1, and a is not 1, so proceed to Step 7.2.

    7.2 a is not 0, so proceed to Step 7.3.

    7.3 $2 = 2^1 * 1$, so $e = 1$, and $a_1 = 1$.

    7.4 e is odd, and $n \equiv 5 \pmod 8$, so set $s = -1$.

    7.5 n is not 3 mod 4, and a_1 is not 3 mod 4, so proceed to step 7.6.

    7.6 $n_1 = 0 = n \bmod 1$.

    7.7   Return $(-1 * \text{Jacobi}(0, 1) = -1)$. This calls Jacobi recursively. Compute the Jacobi symbol for $a = 0$ and $n = 1$:

        7.7.1 $n = 1$, so return 1.

Thus, Jacobi $(0,1) = 1$, so Jacobi $(2,5) = -1*(1) = -1$, and Jacobi $(5, 3439601197) = 1* (-1) = -1$.

C.6   Shawe-Taylor Random_Prime Routine

This routine is recursive and may be used to construct a provable prime number using a hash function.

Let **Hash( )** be the selected hash function, and let *outlen* be the bit length of the hash function output block. The following process or its equivalent **shall** be used to generate a prime number for this constructive method.

ST_Random_Prime ( ):

Input:

1. *length* — The length of the prime to be generated.

2. *input_seed* — The seed to be used for the generation of the requested prime.

Output:

1. *status* — The status returned from the generation routine, where *status* is either **SUCCESS** or **FAILURE.** If **FAILURE** is returned, then zeros are returned as the other output values.

2. *prime* — The requested prime.

3. *prime_seed* — A seed determined during generation.

4. *prime_gen_counter* — (Optional) A counter determined during the generation of the prime.

Process:

1. If (*length* < 2), then return **(FAILURE, 0, 0 {, 0})**.

2. If (*length* ≥ 33), then go to step 14.

3. *prime_seed* = *input_seed*.

4. *prime_gen_counter* = 0.

> Comment: Generate a pseudorandom integer *c* of *length* bits.

5. $c = \textbf{Hash}(\textit{prime\_seed}) \oplus \textbf{Hash}(\textit{prime\_seed} + 1)$.

6. $c = 2^{length-1} + (c \bmod 2^{length-1})$.

7. $c = (2 * \lfloor c / 2 \rfloor) + 1$.

> Comment: Set *prime* to the least odd integer greater than or equal to *c*.

8. *prime_gen_counter* = *prime_gen_counter* + 1.

9. *prime_seed* = *prime_seed* + 2.

10. Perform a deterministic primality test on c. For example, since c is small, its primality can be tested by trial division. See Appendix C.7.

11. If (c is a prime number), then

    11.1   *prime* = c.

    11.2   Return (**SUCCESS**, *prime*, *prime_seed* {, *prime_gen_counter*}).

12. If ($prime\_gen\_counter > (4 * length)$), then return (**FAILURE**, 0, 0 {, 0}).

13. Go to step 5.

14. ($status$, c_0, *prime_seed*, *prime_gen_counter*) = (**ST_Random_Prime** (($\lceil length / 2 \rceil$ + 1), *input_seed*).

15. If **FAILURE** is returned, return (**FAILURE**, 0, 0 {, 0}).

16. $iterations = \lceil length / outlen \rceil - 1$.

17. $old\_counter = prime\_gen\_counter$.

Comment: Generate a pseudorandom integer x in the interval $[2^{length-1}, 2^{length}]$.

18. $x = 0$.

19. For $i = 0$ to *iterations* do

    $x = x + (\textbf{Hash}(prime\_seed + i) * 2^{i \times outlen})$.

20. $prime\_seed = prime\_seed + iterations + 1$.

21. $x = 2^{length-1} + (x \bmod 2^{length-1})$.

Comment: Generate a candidate prime c in the interval $[2^{length-1}, 2^{length}]$.

22. $t = \lceil x / (2c_0) \rceil$.

23. If ($2tc_0 + 1 > 2^{length}$), then $t = \lceil 2^{length-1} / (2c_0) \rceil$.

24. $c = 2tc_0 + 1$.

25. $prime\_gen\_counter = prime\_gen\_counter + 1$.

Comment: Test the candidate prime c for primality; first pick an integer a between 2 and $c - 2$.

26. $a = 0$.

27. For $i = 0$ to *iterations* do

    $a = a + (\textbf{Hash}(prime\_seed + i) * 2^{i * outlen})$.

28. *prime_seed* = *prime_seed* + *iterations* + 1.

29. $a = 2 + (a \bmod (c - 3))$.

30. $z = a^{2t} \bmod c$.

31. If $((1 = \mathbf{GCD}(z - 1, c))$ and $(1 = z^{c_0} \bmod c))$, then

    31.1   *prime* = *c*.

    31.2   Return (**SUCCESS**, *prime*, *prime_seed* {, *prime_gen_counter*}).

32. If $(prime\_gen\_counter \geq ((4 * length) + old\_counter))$, then return (**FAILURE**, 0, 0 {, 0}).

33. $t = t + 1$.

34. Go to step 23.

C.7 Trial Division

An integer is proven to be prime by showing that it has no prime factors less than or equal to its square root. This procedure is not recommended for testing any integers longer than 10 digits.

To prove that *c* is prime:

1. Prepare a table of primes less than $\sqrt{c}$. This can be done by applying the sieve procedure in Appendix C.8.

2. Divide *c* by every prime in the table. If *c* is divisible by one of the primes, then declare that *c* is composite and exit. If convenient, *c* may be divided by composite numbers. For example, rather than preparing a table of primes, it might be more convenient to divide by all integers except those divisible by 3 or 5.

3. Otherwise, declare that *c* is prime and exit.

C.8 Sieve Procedure

A *sieve procedure* is described as follows: Given a sequence of integers $Y_0, Y_0 + 1, \ldots, Y_0 + J$, a sieve will identify the integers in the sequence that are divisible by primes up to some selected limit.

Note that the definitions of the mathematical symbols in this process (e.g., *h*, *L*, *M*, *p*) are internal to this process only, and should not be confused with their use elsewhere in this Standard.

Start by selecting a *factor base* of all the primes p_j, from 2 up to some selected limit *L*. The value of *L* is arbitrary and may be determined by computer limitations. A good, typical value of *L* would be anywhere from 10^3 to 10^5.

1. Compute $S_j = Y_0 \bmod p_j$ for all p_j in the factor base.

2. Initialize an array of length $J + 1$ to zero.

3. Starting at $Y_0 - S_j + p_j$, let every $p_j{}^{th}$ element of the array be set to 1. Do this for the entire length of the array and for every j.

4. When finished, every location in the array that has the value 1 is divisible by some small prime, and is therefore a composite.

The array can be either a bit array for compactness when memory is small, or a byte array for speed when memory is readily available. There is no need to sieve the entire sieve interval at once. The array can be partitioned into suitably small pieces, sieving each piece before going on to the next piece. When finished, every location with the value 0 is a candidate for prime testing.

The amount of work for this procedure is approximately $M \log \log L$, where M is the length of the sieve interval; this is a very efficient procedure for removing composite candidates for primality testing. If $L = 10^5$, the sieve will remove about 96% of all composites.

In some cases, rather than having a set of consecutive integers to sieve, the set of integers to be tested consists of integers lying in an arithmetic progression $Y_0, Y_0 + h, Y_0 + 2h, ..., Y_0 + Jh$, where h is large and not divisible by any primes in the factor base.

1. Select a factor base and initialize an array of length $J + 1$ to 0.

2. Compute $S_j = Y_0 \bmod p_j$ for all p_j in the factor base.

3. Compute $T_j = h \bmod p_j$ and $r = -S_j T_j{}^{-1} \bmod p_j$.

4. Starting at $Y_0 + r$, let every $p_j{}^{th}$ element of the array be set to 1. Do this for the entire length of the array and for every j. Note that the position $Y_0 + r$ in the array actually denotes the number $Y_0 + rh$.

5. When finished, every location in the array that has the value 1 is divisible by some small prime and is therefore composite.

Note: The prime "2" takes the longest amount of time ($M/2$) to sieve, since it touches the most locations in the sieve array. An easy optimization is to combine the initialization of the sieve array with the sieving of the prime "2". It is also possible to sieve the prime "3" during initialization. These optimizations can save about 1/3 of the total sieve time.

C.9   Compute a Probable Prime Factor Based on Auxiliary Primes

This routine constructs a probable prime (a candidate for p or q) using two auxiliary prime numbers and the Chinese Remainder Theorem (CRT).

Input:

r_1 and r_2 Two odd prime numbers satisfying
$\log_2(r_1 r_2) \le (nlen/2) - \log_2(nlen/2) - 6$.

nlen	The desired length of *n*, the RSA modulus.
e	The public verification exponent.
security_strength	The minimum security strength required for random number generation.

Output:

status	The status returned from the generation procedure, where *status* is either **SUCCESS** or **FAILURE**. If **FAILURE** is returned, then zeros are returned as the other output values.
private_prime_factor	The prime factor of *n*.
X	The random number used during the generation of the *private_prime_factor*.

Process:

1. If ($\mathbf{GCD}(2r_1, r_2) \neq 1$), then return (**FAILURE**, 0, 0).

2. $R = ((r_2^{-1} \bmod 2r_1) * r_2) - (((2r_1)^{-1} \bmod r_2) * 2r_1)$.

    Comment: Apply the CRT, so that $R \equiv 1 \pmod{2r_1}$ and $R \equiv -1 \pmod{r_2}$.

3. Generate a random number *X* using an approved random number generator that supports the *security_ strength*, such that $\left(\sqrt{2}\right)\left(2^{nlen/2-1}\right) \le X \le \left(2^{nlen/2} - 1\right)$.

4. $Y = X + ((R - X) \bmod 2r_1 r_2)$.

    Comment: *Y* is the first odd integer $\ge X$, such that r_1 is a prime factor of *Y*–1, and r_2 is a prime factor of *Y*+1.

    Comment: Determine the requested prime number by constructing candidates from a sequence and performing primality tests.

5. $i = 0$.

6. If ($Y \ge 2^{nlen/2}$), then go to step 3.

7. If ($\mathbf{GCD}(Y-1, e) = 1$), then

    7.1   Check the primality of *Y* as specified in Appendix C.3. If **PROBABLY PRIME** is ***not*** returned, go to step 8.

    7.2   *private_prime_factor* = *Y*.

    7.3   Return (**SUCCESS**, *private_prime_factor*, *X*).

8. $i = i + 1$.

9. If $(i \geq 5(nlen/2))$, then return (**FAILURE**, 0, 0).

10. $Y = Y + (2r_1 r_2)$.

11. Go to step 6.

C.10 Construct a Provable Prime (possibly with Conditions), Based on Contemporaneously Constructed Auxiliary Provable Primes

The following process (or its equivalent) **shall** be used to generate an L-bit provable prime p (a candidate for one of the prime factors of an RSA modulus). Note that the use of p in this specification is used generically; both RSA prime factors p and q may be generated using this method.

If a so-called "strong prime" is required, this process can generate primes p_1 and p_2 (of specified bit-lengths N_1 and N_2) that divide $p-1$ and $p+1$, respectively. The resulting prime p will satisfy the conditions traditionally required of a strong prime, provided that the requested bit-lengths for p_1 and p_2 have appropriate sizes.

Regardless of the bit-lengths selected for p_1 and p_2, the quantity $p-1$ will have a prime divisor p_0 whose bit-length is slightly more than half that of p. In addition, the quantity $p_0 - 1$ will have a prime divisor whose bit-length is slightly more than half that of p_0.

This algorithm requires that $N_1 + N_2 \leq L - \lceil L/2 \rceil - 4$. Values for N_1 and N_2 **should** be chosen such that $N_1 + N_2 \leq (L/2) - \log_2(L) - 7$, to ensure that the algorithm can generate as many as $5L$ distinct candidates for p.

Let **Hash** be the selected hash function to be used, and let *outlen* be the bit length of the hash function output block.

Provable_Prime_Construction():

Input:

1.	L	A positive integer equal to the requested bit-length for p. Note that acceptable values for $L = nlen/2$ are computed as specified in Appendix B.3.1, criteria 2(b) and (c), with *nlen* assuming a value specified in Table B.1.
2.	N_1	A positive integer equal to the requested bit-length for p_1. If $N_1 \geq 2$, then p_1 is an odd prime of N_1 bits; otherwise, $p_1 = 1$. Acceptable values for $N_1 \geq 2$ are provided in Table B.1
3.	N_2	A positive integer equal to the requested bit-length for p_2. If $N_2 \geq 2$, then p_2 is an odd prime of N_2 bits; otherwise, $p_2 = 1$. Acceptable values for $N_2 \geq 2$ are provided in Table B.1
4.	*firstseed*	A bit string equal to the first seed to be used.

          The public verification exponent.

Output:

1. *status*   The status returned from the generation procedure, where *status* is either **SUCCESS** or **FAILURE**. If **FAILURE** is returned, then zeros are returned as the other output values.

2. p, p_1, p_2 The required prime p, along with p_1 and p_2 having the property that p_1 divides $p-1$ and p_2 divides $p+1$.

3. *pseed*   A seed determined during generation.

Process:

1. If L, N_1, and N_2 are not acceptable, then, return (**FAILURE**, 0, 0, 0, 0).

   > Comment: Generate p_1 and p_2, as well as the prime p_0.

2. If $N_1 = 1$, then

   2.1 $p_1 = 1$.

   2.2 $p_2seed = firstseed$.

3. If $N_1 \geq 2$, then

   3.1 Using N_1 as the length and *firstseed* as the *input_seed*, use the random prime generation routine in Appendix C.6 to obtain p_1 and p_2seed.

   3.2 If **FAILURE** is returned, then return (**FAILURE**, 0, 0, 0, 0).

4. If $N_2 = 1$, then

   4.1 $p_2 = 1$.

   4.2 $p_0seed = p_2seed$.

5. If $N_2 \geq 2$, then

   5.1 Using N_2 as the length and p_2seed as the *input_seed*, use the random prime generation routine in Appendix C.6 to obtain p_2 and p_0seed.

   5.2 If **FAILURE** is returned, then return (**FAILURE**, 0, 0, 0, 0).

6. Using $\lceil L/2 \rceil + 1$ as the length and p_0seed as the *input_seed*, use the random prime generation routine in Appendix C.6 to obtain p_0 and *pseed*. If **FAILURE** is returned, then return (**FAILURE**, 0, 0, 0, 0).

   > Comment: Generate a (strong) prime p in the interval $[(\sqrt{2})(2^{L-1}), 2^L - 1]$.

7. *iterations* $= \lceil L / outlen \rceil - 1$.

8. *pgen_counter* $= 0$.

> Comment: Generate pseudo-random x in the interval $[(\sqrt{2})(2^{L-1})-1, 2^L-1]$.

9. $x = 0$.

10. For $i = 0$ to *iterations* do
$$x = x + (\mathbf{Hash}(pseed + i)) * 2^{\,i\,*\,outlen}.$$

11. *pseed* $=$ *pseed* $+$ *iterations* $+ 1$.

12. $x = \lfloor (\sqrt{2})(2^{L-1}) \rfloor + (\,x \bmod (2^L - \lfloor (\sqrt{2})(2^{L-1}) \rfloor)\,)$.

> Comment: Generate a candidate for the prime p.

13. If $(\mathbf{GCD}(p_0 p_1, p_2) \neq 1)$, then return $(\mathbf{FAILURE}, 0, 0, 0, 0)$.

14. Compute y in the interval $[1, p_2]$ such that $0 = (y\, p_0\, p_1 - 1) \bmod p_2$.

15. $t = \lceil ((2\,y\, p_0\, p_1) + x)/(2\, p_0\, p_1\, p_2) \rceil$.

16. If $((2(t\, p_2 - y)\, p_0\, p_1 + 1) > 2^L)$, then
$$t = \lceil (\,(2\,y\, p_0\, p_1) + \lfloor (\sqrt{2})(2^{L-1}) \rfloor\,) / (2\, p_0\, p_1\, p_2) \rceil.$$

> Comment: p satisfies
> $0 = (p-1) \bmod (2p_0\, p_1)$ and
> $0 = (p+1) \bmod p_2$.

17. $p = 2(t\, p_2 - y)\, p_0\, p_1 + 1$.

18. *pgen_counter* $=$ *pgen_counter* $+ 1$.

19. If $(\mathbf{GCD}(p-1, e) = 1)$, then

> Comment: Choose an integer a in the interval $[2, p-2]$.

19.1 $a = 0$

19.2 For $i = 0$ to *iterations* do
$$a = a + (\mathbf{Hash}(pseed + i)) * 2^{\,i\,*\,outlen}.$$

19.3 *pseed* $=$ *pseed* $+$ *iterations* $+ 1$.

19.4 $a = 2 + (a \bmod (p-3))$.

> Comment: Test p for primality:

19.5 $z = a^{2(t\, p_2 - y)\, p_1} \bmod p$.

85

19.6　If $((1 = \mathbf{GCD}(z-1, p))$ and $(1 = (z^{p_0} \bmod p))$, then return (**SUCCESS**, p, p_1, p_2, *pseed*).

20. If (*pgen_counter* $\geq 5L$), then return (**FAILURE**, 0, 0, 0, 0).

21. $t = t + 1$.

22. Go to step 16.

Appendix D: Recommended Elliptic Curves for Federal Government Use

This collection of elliptic curves is recommended for Federal government use and contains choices for the private key length and underlying fields. These curves were generated using SHA-1 and the method given in the ANS X9.62 and IEEE Standard 1363-2000 standards. This appendix describes the process that was used. Note that these curves are the same as those included in the previous version of this Standard.

D.1 NIST Recommended Elliptic Curves

D.1.1 Choices

D.1.1.1 Choice of Key Lengths

The principal parameters for elliptic curve cryptography are the elliptic curve E and a designated point G on E called the *base point*. The base point has order n, which is a large prime. The number of points on the curve is hn for some integer h (the *cofactor*), which is not divisible by n. For efficiency reasons, it is desirable to have the cofactor be as small as possible.

All of the curves given below have cofactors 1, 2, or 4. As a result, the private and public keys for a curve are approximately the same length.

D.1.1.2 Choice of Underlying Fields

For each key length, two kinds of fields are provided.

- A *prime field* is the field $GF(p)$, which contains a prime number p of elements. The elements of this field are the integers modulo p, and the field arithmetic is implemented in terms of the arithmetic of integers modulo p.

- A *binary field* is the field $GF(2^m)$, which contains 2^m elements for some m (called the *degree* of the field). The elements of this field are the bit strings of length m, and the field arithmetic is implemented in terms of operations on the bits.

The security strengths for five ranges of the bit length of n is provided in SP 800-57. For the field $GF(p)$, the security strength is dependent on the length of the binary expansion of p. For the field $GF(2^m)$, the security strength is dependent on the value of m. Table E-1 provides the bit lengths of the various underlying fields of the curves provided in this appendix. Column 1 lists the ranges for the bit length of n (also see Table 1 in Section 6.1.1). Column 2 identifies the value of p used for the curves over prime fields, where **len**(p) is the length of the binary expansion of the integer p. Column 3 provides the value of m for the curves over binary fields.

Table D-1: Bit Lengths of the Underlying Fields of the Recommended Curves

Bit Length of n	Prime Field	Binary Field
161 – 223	$\text{len}(p) = 192$	$m = 163$
224 – 255	$\text{len}(p) = 224$	$m = 233$
256 – 383	$\text{len}(p) = 256$	$m = 283$
384 – 511	$\text{len}(p) = 384$	$m = 409$
≥ 512	$\text{len}(p) = 521$	$m = 571$

D.1.1.3 Choice of Basis for Binary Fields

To describe the arithmetic of a binary field, it is first necessary to specify how a bit string is to be interpreted. This is referred to as choosing a *basis* for the field. There are two common types of bases: a *polynomial basis* and a *normal basis*.

- A polynomial basis is specified by an irreducible polynomial modulo 2, called the *field polynomial*. The bit string $(a_{m-1} \ldots a_2\ a_1\ a_0)$ is taken to represent the polynomial

$$a_{m-1} t^{m-1} + \ldots + a_2 t^2 + a_1 t + a_0$$

  over $GF(2)$. The field arithmetic is implemented as polynomial arithmetic modulo $p(t)$, where $p(t)$ is the field polynomial.

- A normal basis is specified by an element θ of a particular kind. The bit string $(a_0\ a_1\ a_2 \ldots a_{m-1})$ is taken to represent the element

$$a_0 \theta + a_1 \theta^2 + a_2 \theta^{2^2} + \ldots + a_{m-1} \theta^{2^{m-1}}.$$

  Normal basis field arithmetic is not easy to describe or efficient to implement in general, except for a special class called *Type T low-complexity* normal bases. For a given field degree m, the choice of T specifies the basis and the field arithmetic (see Appendix D.3).

There are many polynomial bases and normal bases from which to choose. The following procedures are commonly used to select a basis representation.

- *Polynomial Basis*: If an irreducible *trinomial* $t^m + t^k + 1$ exists over $GF(2)$, then the field polynomial $p(t)$ is chosen to be the irreducible trinomial with the lowest-degree middle term t^k. If no irreducible trinomial exists, then a *pentanomial* $t^m + t^a + t^b + t^c + 1$ is selected. The particular pentanomial chosen has the following properties: the second term t^a has the lowest degree m; the third term t^b has the lowest degree among all irreducible pentanomials of degree m and second term t^a; and the fourth term t^c has the lowest degree among all irreducible pentanomials of degree m, second term t^a, and third term t^b.

- *Normal Basis*: Choose the Type T low-complexity normal basis with the smallest *T*.

For each binary field, the parameters are given for the above basis representations.

D.1.1.4 Choice of Curves

Two kinds of curves are given:

- *Pseudo-random* curves are those whose coefficients are generated from the output of a seeded cryptographic hash function. If the domain parameter seed value is given along with the coefficients, it can be easily verified that the coefficients were generated by that method.

- *Special curves* are those whose coefficients and underlying field have been selected to optimize the efficiency of the elliptic curve operations.

For each curve size range, the following curves are given:

→ A pseudo-random curve over $GF(p)$.

→ A pseudo-random curve over $GF(2^m)$.

→ A special curve over $GF(2^m)$ called a *Koblitz curve* or *anomalous binary curve*.

The pseudo-random curves were generated as specified in ANS X9.62 using SHA-1.

D.1.1.5 Choice of Base Points

Any point of order *n* can serve as the base point. Each curve is supplied with a sample base point $G = (G_x, G_y)$. Users may want to generate their own base points to ensure cryptographic separation of networks. See ANS X9.62 or IEEE Standard 1363-2000.

D.1.2 Curves over Prime Fields

For each prime *p*, a pseudo-random curve

$$E: y^2 \equiv x^3 - 3x + b \pmod{p}$$

of prime order *n* is listed[4]. (Thus, for these curves, the cofactor is always $h = 1$.) The following parameters are given:

- The prime modulus *p*
- The order *n*
- The 160-bit input seed *SEED* to the SHA-1 based algorithm (i.e., the domain parameter seed)
- The output *c* of the SHA-1 based algorithm

---

[4] The selection $a \equiv -3$ for the coefficient of *x* was made for reasons of efficiency; see IEEE Std 1363-2000.

- The coefficient b (satisfying $b^2 c \equiv -27 \pmod p$)
- The base point x coordinate G_x
- The base point y coordinate G_y

The integers p and n are given in decimal form; bit strings and field elements are given in hexadecimal.

D.1.2.1 Curve P-192

$p =$ 6277101735386680763835789423207666416083908700390324961279

$n =$ 6277101735386680763835789423176059013767194773182842284081

$SEED =$ 3045ae6f c8422f64 ed579528 d38120ea e12196d5

$c =$ 3099d2bb bfcb2538 542dcd5f b078b6ef 5f3d6fe2 c745de65

$b =$ 64210519 e59c80e7 0fa7e9ab 72243049 feb8deec c146b9b1

$G_x =$ 188da80e b03090f6 7cbf20eb 43a18800 f4ff0afd 82ff1012

$G_y =$ 07192b95 ffc8da78 631011ed 6b24cdd5 73f977a1 1e794811

D.1.2.2 Curve P-224

$p =$ 2695994666715063979466701508701963067355791626002630814351
0066298881

$n =$ 2695994666715063979466701508701962594045780771442439172168
2722368061

$SEED =$ bd713447 99d5c7fc dc45b59f a3b9ab8f 6a948bc5

$c =$ 5b056c7e 11dd68f4 0469ee7f 3c7a7d74 f7d12111 6506d031
218291fb

$b =$ b4050a85 0c04b3ab f5413256 5044b0b7 d7bfd8ba 270b3943
2355ffb4

$G_x =$ b70e0cbd 6bb4bf7f 321390b9 4a03c1d3 56c21122 343280d6
115c1d21

$G_y =$ bd376388 b5f723fb 4c22dfe6 cd4375a0 5a074764 44d58199
85007e34

90

D.1.2.3 Curve P-256

p = 115792089210356248762697446949407573530086143415290314195533631308867097853951

n = 115792089210356248762697446949407573529996955224135760342422259061068512044369

$SEED$ = c49d3608 86e70493 6a6678e1 139d26b7 819f7e90

c = 7efba166 2985be94 03cb055c 75d4f7e0 ce8d84a9 c5114abc
af317768 0104fa0d

b = 5ac635d8 aa3a93e7 b3ebbd55 769886bc 651d06b0 cc53b0f6
3bce3c3e 27d2604b

G_x = 6b17d1f2 e12c4247 f8bce6e5 63a440f2 77037d81 2deb33a0
f4a13945 d898c296

G_y = 4fe342e2 fe1a7f9b 8ee7eb4a 7c0f9e16 2bce3357 6b315ece
cbb64068 37bf51f5

D.1.2.4 Curve P-384

p = 39402006196394479212279040100143613805079739270465446667948293404245721771496870329047266088258938001861606973112319

n = 39402006196394479212279040100143613805079739270465446667946905279627659399113263569398956308152294913554433653942643

$SEED$ = a335926a a319a27a 1d00896a 6773a482 7acdac73

c = 79d1e655 f868f02f ff48dcde e14151dd b80643c1 406d0ca1
0dfe6fc5 2009540a 495e8042 ea5f744f 6e184667 cc722483

b = b3312fa7 e23ee7e4 988e056b e3f82d19 181d9c6e fe814112
0314088f 5013875a c656398d 8a2ed19d 2a85c8ed d3ec2aef

G_x = aa87ca22 be8b0537 8eb1c71e f320ad74 6e1d3b62 8ba79b98
59f741e0 82542a38 5502f25d bf55296c 3a545e38 72760ab7

G_y = 3617de4a 96262c6f 5d9e98bf 9292dc29 f8f41dbd 289a147c
e9da3113 b5f0b8c0 0a60b1ce 1d7e819d 7a431d7c 90ea0e5f

D.1.2.5 Curve P-521

$p =$ 686479766013060971498190079908139321726943530014330540939
44634591855431833976560521225596406614545549772963113914808580371219879997166438125740282911150571 51

$n =$ 686479766013060971498190079908139321726943530014330540939
44634591855431833976553942450577463332171975329639963713633211138647686124403803403728088927070 05449

$SEED =$ d09e8800 291cb853 96cc6717 393284aa a0da64ba

$c =$ 0b4 8bfa5f42 0a349495 39d2bdfc 264eeeeb 077688e4
4fbf0ad8 f6d0edb3 7bd6b533 28100051 8e19f1b9 ffbe0fe9
ed8a3c22 00b8f875 e523868c 70c1e5bf 55bad637

$b =$ 051 953eb961 8e1c9a1f 929a21a0 b68540ee a2da725b
99b315f3 b8b48991 8ef109e1 56193951 ec7e937b 1652c0bd
3bb1bf07 3573df88 3d2c34f1 ef451fd4 6b503f00

$G_x =$ c6 858e06b7 0404e9cd 9e3ecb66 2395b442 9c648139
053fb521 f828af60 6b4d3dba a14b5e77 efe75928 fe1dc127
a2ffa8de 3348b3c1 856a429b f97e7e31 c2e5bd66

$G_y =$ 118 39296a78 9a3bc004 5c8a5fb4 2c7d1bd9 98f54449
579b4468 17afbd17 273e662c 97ee7299 5ef42640 c550b901
3fad0761 353c7086 a272c240 88be9476 9fd16650

D.1.3 Curves over Binary Fields

For each field degree m, a pseudo-random curve is given, along with a Koblitz curve. The pseudo-random curve has the form

$$E: y^2 + x y = x^3 + x^2 + b,$$

and the Koblitz curve has the form

$$E_a: y^2 + x y = x^3 + a x^2 + 1,$$

where $a = 0$ or 1.

For each pseudorandom curve, the cofactor is $h = 2$. The cofactor of each Koblitz curve is $h = 2$ if $a = 1$, and $h = 4$ if $a = 0$.

92

The coefficients of the pseudo-random curves, and the coordinates of the base points of both kinds of curves, are given in terms of both the polynomial and normal basis representations discussed in Appendix D.1.1.3.

For each m, the following parameters are given:

Field Representation:

- The normal basis type T
- The field polynomial (a trinomial or pentanomial)

Koblitz Curve:

- The coefficient a
- The base point order n
- The base point x coordinate G_x
- The base point y coordinate G_y

Pseudo-random curve:

- The base point order n

Pseudo-random curve (Polynomial Basis representation):

- The coefficient b
- The base point x coordinate G_x
- The base point y coordinate G_y

Pseudo-random curve (Normal Basis representation):

- The 160-bit input seed *SEED* to the SHA-1 based algorithm (i.e., the domain parameter seed)
- The coefficient b (i.e., the output of the SHA-1 based algorithm)
- The base point x coordinate G_x
- The base point y coordinate G_y

Integers (such as T, m, and n) are given in decimal form; bit strings and field elements are given in hexadecimal.

D.1.3.1 Degree 163 Binary Field

$T = \quad 4$

$p(t) = \ t^{163} + t^7 + t^6 + t^3 + 1$

93

D.1.3.1.1 Curve K-163

$a =$ 1

$n =$ 5846006549323611672814741753598448348329118574063

Polynomial Basis:

$G_x =$ 2 fe13c053 7bbc11ac aa07d793 de4e6d5e 5c94eee8

$G_y =$ 2 89070fb0 5d38ff58 321f2e80 0536d538 ccdaa3d9

Normal Basis:

$G_x =$ 0 5679b353 caa46825 fea2d371 3ba450da 0c2a4541

$G_y =$ 2 35b7c671 00506899 06bac3d9 dec76a83 5591edb2

D.1.3.1.2 Curve B-163

$n =$ 5846006549323611672814742442876390689256843201587

Polynomial Basis:

$b =$ 2 0a601907 b8c953ca 1481eb10 512f7874 4a3205fd

$G_x =$ 3 f0eba162 86a2d57e a0991168 d4994637 e8343e36

$G_y =$ 0 d51fbc6c 71a0094f a2cdd545 b11c5c0c 797324f1

Normal Basis:

$SEED =$ 85e25bfe 5c86226c db12016f 7553f9d0 e693a268

$b =$ 6 645f3cac f1638e13 9c6cd13e f61734fb c9e3d9fb

$G_x =$ 0 311103c1 7167564a ce77ccb0 9c681f88 6ba54ee8

$G_y =$ 3 33ac13c6 447f2e67 613bf700 9daf98c8 7bb50c7f

D.1.3.2 Degree 233 Binary Field

$T =$ 2

$p(t) =$ $t^{233} + t^{74} + 1$

D.1.3.2.1 Curve K-233

$a =$ 0

$n =$ 3450873173395281893717377931138512760570940988862252 12\
6328087024741343

Polynomial Basis:

$G_x =$ 172 32ba853a 7e731af1 29f22ff4 149563a4 19c26bf5
0a4c9d6e efad6126

$G_y =$ 1db 537dece8 19b7f70f 555a67c4 27a8cd9b f18aeb9b
56e0c110 56fae6a3

Normal Basis:

$G_x =$ 0fd e76d9dcd 26e643ac 26f1aa90 1aa12978 4b71fc07
22b2d056 14d650b3

$G_y =$ 064 3e317633 155c9e04 47ba8020 a3c43177 450ee036
d6335014 34cac978

D.1.3.2.2 Curve B-233

$n =$ 6901746346790563787434755862277025555838912737345013 55\
5379383634485463

Polynomial Basis:

$b =$ 066 647ede6c 332c7f8c 0923bb58 213b333b 20e9ce42
81fe115f 7d8f90ad

$G_x =$ 0fa c9dfcbac 8313bb21 39f1bb75 5fef65bc 391f8b36
f8f8eb73 71fd558b

$G_y =$ 100 6a08a419 03350678 e58528be bf8a0bef f867a7ca
36716f7e 01f81052

Normal Basis:

$SEED =$ 74d59ff0 7f6b413d 0ea14b34 4b20a2db 049b50c3

$b =$ 1a0 03e0962d 4f9a8e40 7c904a95 38163adb 82521260 0c7752ad 52233279

$G_x =$ 18b 863524b3 cdfefb94 f2784e0b 116faac5 4404bc91 62a363ba b84a14c5

$G_y =$ 049 25df77bd 8b8ff1a5 ff519417 822bfedf 2bbd7526 44292c98 c7af6e02

D.1.3.3   Degree 283 Binary Field

$T =$ 6

$p(t) =$ $t^{283} + t^{12} + t^7 + t^5 + 1$

D.1.3.3.1   Curve K-283

$a =$ 0

$n =$ 3885337784451458141838923813647037813284811733793061324295874997529815829704422603873

Polynomial Basis:

$G_x =$ 503213f 78ca4488 3f1a3b81 62f188e5 53cd265f 23c1567a 16876913 b0c2ac24 58492836

$G_y =$ 1ccda38 0f1c9e31 8d90f95d 07e5426f e87e45c0 e8184698 e4596236 4e341161 77dd2259

Normal Basis:

$G_x =$ 3ab9593 f8db09fc 188f1d7c 4ac9fcc3 e57fcd3b db15024b 212c7022 9de5fcd9 2eb0ea60

$G_y =$ 2118c47 55e7345c d8f603ef 93b98b10 6fe8854f feb9a3b3 04634cc8 3a0e759f 0c2686b1

D.1.3.3.2 Curve B-283

$n =$ 7770675568902916283677847627294075626569625924376904889

109196526770044277787378692871

Polynomial Basis:

$b =$ 27b680a c8b8596d a5a4af8a 19a0303f ca97fd76 45309fa2
a581485a f6263e31 3b79a2f5

$G_x =$ 5f93925 8db7dd90 e1934f8c 70b0dfec 2eed25b8 557eac9c
80e2e198 f8cdbecd 86b12053

$G_y =$ 3676854 fe24141c b98fe6d4 b20d02b4 516ff702 350eddb0
826779c8 13f0df45 be8112f4

Normal Basis:

$SEED =$ 77e2b073 70eb0f83 2a6dd5b6 2dfc88cd 06bb84be

$b =$ 157261b 894739fb 5a13503f 55f0b3f1 0c560116 66331022
01138cc1 80c0206b dafbc951

$G_x =$ 749468e 464ee468 634b21f7 f61cb700 701817e6 bc36a236
4cb8906e 940948ea a463c35d

$G_y =$ 62968bd 3b489ac5 c9b859da 68475c31 5bafcdc4 ccd0dc90
5b70f624 46f49c05 2f49c08c

D.1.3.4 Degree 409 Binary Field

$T =$ 4

$p(t) =$ $t^{409} + t^{87} + 1$

D.1.3.4.1 Curve K-409

$a =$ 0

$n =$ 330527984395124299475957654016385519914202341482140609 64\

232439502288071128924919105067325845777745801409636659061

7731358671

Polynomial Basis:

$G_x =$ 060f05f 658f49c1 ad3ab189 0f718421 0efd0987 e307c84c
27accfb8 f9f67cc2 c460189e b5aaaa62 ee222eb1 b35540cf
e9023746

$G_y =$ 1e36905 0b7c4e42 acba1dac bf04299c 3460782f 918ea427
e6325165 e9ea10e3 da5f6c42 e9c55215 aa9ca27a 5863ec48
d8e0286b

Normal Basis:

$G_x =$ 1b559c7 cba2422e 3affe133 43e808b5 5e012d72 6ca0b7e6
a63aeafb c1e3a98e 10ca0fcf 98350c3b 7f89a975 4a8e1dc0
713cec4a

$G_y =$ 16d8c42 052f07e7 713e7490 eff318ba 1abd6fef 8a5433c8
94b24f5c 817aeb79 852496fb ee803a47 bc8a2038 78ebf1c4
99afd7d6

D.1.3.4.2 Curve B-409

$n =$ 661055968790248598951915308032771039828404682964281219284648798304157774827374805208143723762179110965979867288366567526771

Polynomial Basis:

$b =$ 021a5c2 c8ee9feb 5c4b9a75 3b7b476b 7fd6422e f1f3dd67
4761fa99 d6ac27c8 a9a197b2 72822f6c d57a55aa 4f50ae31
7b13545f

$G_x =$ 15d4860 d088ddb3 496b0c60 64756260 441cde4a f1771d4d
b01ffe5b 34e59703 dc255a86 8a118051 5603aeab 60794e54
bb7996a7

98

$G_y =$ 061b1cf ab6be5f3 2bbfa783 24ed106a 7636b9c5 a7bd198d
0158aa4f 5488d08f 38514f1f df4b4f40 d2181b36 81c364ba
0273c706

Normal Basis:

$SEED =$ 4099b5a4 57f9d69f 79213d09 4c4bcd4d 4262210b

$b =$ 124d065 1c3d3772 f7f5a1fe 6e715559 e2129bdf a04d52f7
b6ac7c53 2cf0ed06 f610072d 88ad2fdc c50c6fde 72843670
f8b3742a

$G_x =$ 0ceacbc 9f475767 d8e69f3b 5dfab398 13685262 bcacf22b
84c7b6dd 981899e7 318c96f0 761f77c6 02c016ce d7c548de
830d708f

$G_y =$ 199d64b a8f089c6 db0e0b61 e80bb959 34afd0ca f2e8be76
d1c5e9af fc7476df 49142691 ad303902 88aa09bc c59c1573
aa3c009a

D.1.3.5 Degree 571 Binary Field

$T =$ 10

$p(t) = t^{571} + t^{10} + t^5 + t^2 + 1$

D.1.3.5.1 Curve K-571

$a =$ 0

$n =$ 1932268761508629172347675945465993672149463664853217499
3286176257257595711447802122681339785227067118347067 1280
0825351461273674974066617311929682421617092503555733 6852
76673

Polynomial Basis:

$G_x =$ 26eb7a8 59923fbc 82189631 f8103fe4 ac9ca297 0012d5d4
60248048 01841ca4 43709584 93b205e6 47da304d b4ceb08c

99

```
          bbd1ba39  494776fb  988b4717  4dca88c7  e2945283  a01c8972
```

$G_y =$
```
          349dc80  7f4fbf37  4f4aeade  3bca9531  4dd58cec  9f307a54
          ffc61efc  006d8a2c  9d4979c0  ac44aea7  4fbebbb9  f772aedc
          b620b01a  7ba7af1b  320430c8  591984f6  01cd4c14  3ef1c7a3
```

Normal Basis:

$G_x =$
```
          04bb2db  a418d0db  107adae0  03427e5d  7cc139ac  b465e593
          4f0bea2a  b2f3622b  c29b3d5b  9aa7a1fd  fd5d8be6  6057c100
          8e71e484  bcd98f22  bf847642  37673674  29ef2ec5  bc3ebcf7
```

$G_y =$
```
          44cbb57  de20788d  2c952d7b  56cf39bd  3e89b189  84bd124e
          751ceff4  369dd8da  c6a59e6e  745df44d  8220ce22  aa2c852c
          fcbbef49  ebaa98bd  2483e331  80e04286  feaa2530  50caff60
```

D.1.3.5.2 Curve B-571

$n =$
```
          3864537523017258344695351890931987344298927329706434998
          6572352514515191422895604245361439993894157730831338112
          1926944486246872462816813070234528288303332411393191105 2
          85703
```

Polynomial Basis:

$b =$
```
          2f40e7e  2221f295  de297117  b7f3d62f  5c6a97ff  cb8ceff1
          cd6ba8ce  4a9a18ad  84ffabbd  8efa5933  2be7ad67  56a66e29
          4afd185a  78ff12aa  520e4de7  39baca0c  7ffeff7f  2955727a
```

$G_x =$
```
          303001d  34b85629  6c16c0d4  0d3cd775  0a93d1d2  955fa80a
          a5f40fc8  db7b2abd  bde53950  f4c0d293  cdd711a3  5b67fb14
          99ae6003  8614f139  4abfa3b4  c850d927  e1e7769c  8eec2d19
```

$G_y =$
```
          37bf273  42da639b  6dccfffe  b73d69d7  8c6c27a6  009cbbca
          1980f853  3921e8a6  84423e43  bab08a57  6291af8f  461bb2a8
          b3531d2f  0485c19b  16e2f151  6e23dd3c  1a4827af  1b8ac15b
```

100

Normal Basis:

$SEED =$ 2aa058f7 3a0e33ab 486b0f61 0410c53a 7f132310

$b =$ 3762d0d 47116006 179da356 88eeaccf 591a5cde a7500011
          8d9608c5 9132d434 26101a1d fb377411 5f586623 f75f0000
          1ce61198 3c1275fa 31f5bc9f 4be1a0f4 67f01ca8 85c74777

$G_x =$ 0735e03 5def5925 cc33173e b2a8ce77 67522b46 6d278b65
          0a291612 7dfea9d2 d361089f 0a7a0247 a184e1c7 0d417866
          e0fe0feb 0ff8f2f3 f9176418 f97d117e 624e2015 df1662a8

$G_y =$ 04a3642 0572616c df7e606f ccadaecf c3b76dab 0eb1248d
          d03fbdfc 9cd3242c 4726be57 9855e812 de7ec5c5 00b4576a
          24628048 b6a72d88 0062eed0 dd34b109 6d3acbb6 b01a4a97

D.2   Implementation of Modular Arithmetic

The prime moduli in the above examples are of a special type (called *generalized Mersenne numbers*) for which modular multiplication can be carried out more efficiently than in general. This section provides the rules for implementing this faster arithmetic for each of the prime moduli appearing in the examples.

The usual way to multiply two integers (mod m) is to take the integer product and reduce it (mod m). One therefore has the following problem: given an integer A less than m^2, compute

$$B = A \bmod m.$$

In general, one must obtain B as the remainder of an integer division. If m is a generalized Mersenne number, however, then B can be expressed as a sum or difference (mod m) of a small number of terms. To compute this expression, the integer sum or difference can be evaluated and the result reduced modulo m. The latter reduction can be accomplished by adding or subtracting a few copies of m.

The prime modulus p for each of the five example curves is a generalized Mersenne number.

D.2.1 Curve P-192

The modulus for this curve is $p = 2^{192} - 2^{64} - 1$. Every integer A less than p^2 can be written as

$$A = A_5 \cdot 2^{320} + A_4 \cdot 2^{256} + A_3 \cdot 2^{192} + A_2 \cdot 2^{128} + A_1 \cdot 2^{64} + A_0,$$

where each A_i is a 64-bit integer. As a concatenation of 64-bit words, this can be denoted by

$$A = (A_5 \, \| \, A_4 \, \| \, A_3 \, \| \, A_2 \, \| \, A_0).$$

The expression for B is

$$B = T + S_1 + S_2 + S_3 \bmod p,$$

where the 192-bit terms are given by

$$T = (A_2 \| A_1 \| A_0)$$
$$S_1 = (A_3 \| A_3)$$
$$S_2 = (A_4 \| A_4 \| 0)$$
$$S_3 = (A_5 \| A_5 \| A_5).$$

D.2.2 Curve P-224

The modulus for this curve is $p = 2^{224} - 2^{96} + 1$. Every integer A less than p^2 can be written as:

$$A = A_{13} \cdot 2^{416} + A_{12} \cdot 2^{384} + A_{11} \cdot 2^{352} + A_{10} \cdot 2^{320} + A_9 \cdot 2^{288} + A_8 \cdot 2^{256} + A_7 \cdot 2^{224} + A_6 \cdot 2^{192} +$$
$$A_5 \cdot 2^{160} + A_4 \cdot 2^{128} + A_3 \cdot 2^{96} + A_2 \cdot 2^{64} + A_1 \cdot 2^{32} + A_0,$$

where each A_i is a 32-bit integer. As a concatenation of 32-bit words, this can be denoted by:

$$A = (A_{13} \| A_{12} \| \ldots \| A_0).$$

The expression for B is:

$$B = T + S_1 + S_2 - D_1 - D_2 \bmod p,$$

where the 224-bit terms are given by:

$$T = (A_6 \| A_5 \| A_4 \| A_3 \| A_2 \| A_1 \| A_0)$$
$$S_1 = (A_{10} \| A_9 \| A_8 \| A_7 \| 0 \| 0 \| 0)$$
$$S_2 = (0 \| A_{13} \| A_{12} \| A_{11} \| 0 \| 0 \| 0)$$
$$D_1 = (A_{13} \| A_{12} \| A_{11} \| A_{10} \| A_9 \| A_8 \| A_7)$$
$$D_2 = (0 \| 0 \| 0 \| 0 \| A_{13} \| A_{12} \| A_{11}).$$

D.2.3 Curve P-256

The modulus for this curve is $p = 2^{256} - 2^{224} + 2^{192} + 2^{96} - 1$. Every integer A less than p^2 can be written as:

$$A = A_{15} \cdot 2^{480} + A_{14} \cdot 2^{448} + A_{13} \cdot 2^{416} + A_{12} \cdot 2^{384} + A_{11} \cdot 2^{352} + A_{10} \cdot 2^{320} + A_9 \cdot 2^{288} + A_8 \cdot 2^{256} +$$
$$A_7 \cdot 2^{224} + A_6 \cdot 2^{192} + A_5 \cdot 2^{160} + A_4 \cdot 2^{128} + A_3 \cdot 2^{96} + A_2 \cdot 2^{64} + A_1 \cdot 2^{32} + A_0,$$

where each A_i is a 32-bit integer. As a concatenation of 32-bit words, this can be denoted by

$$A = (A_{15} \| A_{14} \| \cdots \| A_0).$$

The expression for B is:

$$B = T + 2S_1 + 2S_2 + S_3 + S_4 - D_1 - D_2 - D_3 - D_4 \bmod p,$$

where the 256-bit terms are given by:

$T = (A_7 \| A_6 \| A_5 \square \| A_4 \| A_3 \| A_2 \| A_1 \| A_0)$

$S_1 = (A_{15} \| A_{14} \square \| A_{13} \| A_{12} \| A_{11} \| 0 \| 0 \| 0)$

$S_2 = (0 \| A_{15} \| A_{14} \square \| A_{13} \| A_{12} \| 0 \| 0 \| 0)$

$S_3 = (A_{15} \| A_{14} \square \| 0 \| 0 \| 0 \| A_{10} \| A_9 \| A_8)$

$S_4 = (A_8 \| A_{13} \| A_{15} \| A_{14} \| A_{13} \| A_{11} \| A_{10} \| A_9)$

$D_1 = (A_{10} \| A_8 \| 0 \| 0 \| 0 \| A_{13} \| A_{12} \| A_{11})$

$D_2 = (A_{11} \| A_9 \| 0 \| 0 \| A_{15} \| A_{14} \| A_{13} \| A_{12} \square)$

$D_3 = (A_{12} \square \| 0 \| A_{10} \| A_9 \| A_8 \| A_{15} \| A_{14} \| A_{13})$

$D_4 = (A_{13} \| 0 \| A_{11} \| A_{10} \| A_9 \| 0 \| A_{15} \| A_{14})$

D.2.4 Curve P-384

The modulus for this curve is $p = 2^{384} - 2^{128} - 2^{96} + 2^{32} - 1$. Every integer A less than p^2 can be written as:

$$A = A_{23} \cdot 2^{736} + A_{22} \cdot 2^{704} + A_{21} \cdot 2^{672} + A_{20} \cdot 2^{640} + A_{19} \cdot 2^{608} + A_{18} \cdot 2^{576} + A_{17} \cdot 2^{544} + A_{16} \cdot 2^{512} +$$
$$A_{15} \cdot 2^{480} + A_{14} \cdot 2^{448} + A_{13} \cdot 2^{416} + A_{12} \cdot 2^{384} + A_{11} \cdot 2^{352} + A_{10} \cdot 2^{320} + A_9 \cdot 2^{288} + A_8 \cdot 2^{256} +$$
$$A_7 \cdot 2^{224} + A_6 \cdot 2^{192} + A_5 \cdot 2^{160} + A_4 \cdot 2^{128} + A_3 \cdot 2^{96} + A_2 \cdot 2^{64} + A_1 \cdot 2^{32} + A_0,$$

where each A_i is a 32-bit integer. As a concatenation of 32-bit words, this can be denoted by

$$A = (A_{23} \| A_{22} \| \cdots \| A_0).$$

The expression for B is:

$$B = T + 2S_1 + S_2 + S_3 + S_4 + S_5 + S_6 - D_1 - D_2 - D_3 \bmod p,$$

where the 384-bit terms are given by:

$T = \quad (A_{11} \| A_{10} \| A_9 \| A_8 \| A_7 \| A_6 \| A_5 \| A_4 \| A_3 \| A_2 \| A_1 \| A_0)$

$S_1 = \quad (0 \| 0 \| 0 \| 0 \| 0 \| A_{23} \| A_{22} \| A_{21} \| 0 \| 0 \| 0 \| 0)$

$S_2 = \quad (A_{23} \| A_{22} \| A_{21} \| A_{20} \| A_{19} \| A_{18} \| A_{17} \| A_{16} \| A_{15} \| A_{14} \| A_{13} \| A_{12})$

$S_3 = \quad (A_{20} \| A_{19} \| A_{18} \| A_{17} \| A_{16} \| A_{15} \| A_{14} \| A_{13} \| A_{12} \| A_{23} \| A_{22} \| A_{21})$

$S_4 = \quad (A_{19} \| A_{18} \| A_{17} \| A_{16} \| A_{15} \| A_{14} \| A_{13} \| A_{12} \| A_{20} \| 0 \| A_{23} \| 0)$

$S_5 = \quad (0 \| 0 \| 0 \| 0 \| A_{23} \| A_{22} \| A_{21} \| A_{20} \| 0 \| 0 \| 0 \| 0)$

$$S_6 = (\,0 \parallel 0 \parallel 0 \parallel 0 \parallel 0 \parallel 0 \parallel A_{23} \parallel A_{22} \parallel A_{21} \parallel 0 \parallel 0 \parallel A_{20}\,)$$

$$D_1 = (A_{22} \parallel A_{21} \parallel A_{20} \parallel A_{19} \parallel A_{18} \parallel A_{17} \parallel A_{16} \parallel A_{15} \parallel A_{14} \parallel A_{13} \parallel A_{12} \parallel A_{23}\,)$$

$$D_2 = (\,0 \parallel 0 \parallel 0 \parallel 0 \parallel 0 \parallel 0 \parallel 0 \parallel A_{23} \parallel A_{22} \parallel A_{21} \parallel A_{20} \parallel 0\,)$$

$$D_3 = (\,0 \parallel 0 \parallel 0 \parallel 0 \parallel 0 \parallel 0 \parallel 0 \parallel A_{23} \parallel A_{23} \parallel 0 \parallel 0 \parallel 0\,).$$

D.2.5 Curve P-521

The modulus for this curve is $p = 2^{521} - 1$. Every integer A less than p^2 can be written

$$A = A_1 \cdot 2^{521} + A_0,$$

where each A_i is a 521-bit integer. As a concatenation of 521-bit words, this can be denoted by

$$A = (A_1 \parallel A_0).$$

The expression for B is:

$$B = (A_0 + A_1) \bmod p.$$

D.3 Normal Bases

The elements of $GF(2^m)$ are expressed in terms of the type T normal *basis*[5] B for $GF(2^m)$, for some T. Each element has a unique representation as a bit string:

$$(\,a_0\, a_1\, \ldots\, a_{m-1}\,).$$

The arithmetic operations are performed as follows.

Addition: addition of two elements is implemented by bit-wise addition modulo 2. Thus, for example,

$$(1100111) + (1010010) = (0110101).$$

Squaring: if

$$\alpha = (\,a_0\, a_1\, \ldots\, a_{m-1}\,)$$

then

$$\alpha^2 = (a_{m-1}\, a_0\, a_1\, \ldots\, a_{m-2}\,).$$

Multiplication: to perform multiplication, a function $F(\underline{u}, \underline{v})$ is constructed on inputs

---

[5] It is assumed in this section that m is odd and T is even, since this is the only case considered in this Standard.

$$\underline{u} = (\, u_0 \, u_1 \, \ldots \, u_{m-1} \,) \quad \text{and} \quad \underline{v} = (\, v_0 \, v_1 \, \ldots \, v_{m-1} \,)$$

as follows.

1. Set $p \leftarrow Tm + 1$.

2. Let u be an integer having order T modulo p.

3. Compute the sequence $F(1), F(2), \ldots, F(p-1)$ as follows:

    3.1   Set $w \leftarrow 1$.

    3.2   For j from 0 to $T-1$ do

        3.2.1   Set $n \leftarrow w$.

        3.2.2   For $i = 0$ to $m-1$ do

            3.2.2.1   Set $F(n) \leftarrow i$.

            3.2.2.2   Set $n \leftarrow 2n \bmod p$.

        3.2.3   Set $w \leftarrow uw \bmod p$.

4. Output the formula:

$$F(u,v) := \sum_{k=1}^{p-2} u_{F(k+1)} v_{F(p-k)}.$$

This computation need only be performed once per basis.

Given the function F for B, the product

$$(\, c_0 \, c_1 \, \ldots \, c_{m-1} \,) = (\, a_0 \, a_1 \, \ldots \, a_{m-1} \,) * (\, b_0 \, b_1 \, \ldots \, b_{m-1} \,)$$

is computed as follows:

1. Set $(\, u_0 \, u_1 \, \ldots \, u_{m-1} \,) \leftarrow (\, a_0 \, a_1 \, \ldots \, a_{m-1} \,)$.

2. Set $(\, v_0 \, v_1 \, \ldots \, v_{m-1} \,) \leftarrow (\, b_0 \, b_1 \, \ldots \, b_{m-1} \,)$.

3. For $k = 0$ to $m - 1$ do

    3.1   Compute

$$c_k = F(\underline{u}, \underline{v}).$$

    3.2   Set $u \leftarrow$ **LeftShift** (u) and $v \leftarrow$ **LeftShift** (v), where **LeftShift** denotes the circular left shift operation.

4. Output $c = (\, c_0 \, c_1 \, \ldots \, c_{m-1} \,)$.

Example: For the type 4 normal basis for $GF(2^7)$, $p = 29$ and $u = 12$ or 17. Thus, the values of F are given by:

$F(1) = 0$ $F(8) = 3$ $F(15) = 6$ $F(22) = 5$

$F(2) = 1$ $F(9) = 3$ $F(16) = 4$ $F(23) = 6$

$F(3) = 5$ $F(10) = 2$ $F(17) = 0$ $F(24) = 1$

$F(4) = 2$ $F(11) = 4$ $F(18) = 4$ $F(25) = 2$

$F(5) = 1$ $F(12) = 0$ $F(19) = 2$ $F(26) = 5$

$F(6) = 6$ $F(13) = 4$ $F(20) = 3$ $F(27) = 1$

$F(7) = 5$ $F(14) = 6$ $F(21) = 3$ $F(28) = 0$

Therefore,

$$F(\underline{u}, \underline{v}) = u_0 v_1 + u_1 (v_0 + v_2 + v_5 + v_6) + u_2 (v_1 + v_3 + v_4 + v_5) + u_3 (v_2 + v_5) +$$
$$u_4 (v_2 + v_6) + u_5 (v_1 + v_2 + v_3 + v_6) + u_6 (v_1 + v_4 + v_5 + v_6).$$

Thus, if

$$a = (1\ 0\ 1\ 0\ 1\ 1\ 1) \text{ and } b = (1\ 1\ 0\ 0\ 0\ 0\ 1),$$

then

$$c_0 = F((1\ 0\ 1\ 0\ 1\ 1\ 1), (1\ 1\ 0\ 0\ 0\ 0\ 1)) = 1,$$
$$c_1 = F((0\ 1\ 0\ 1\ 1\ 1\ 1), (1\ 0\ 0\ 0\ 0\ 1\ 1)) = 0,$$
$$\vdots$$
$$c_6 = F((1\ 1\ 0\ 1\ 0\ 1\ 1), (1\ 1\ 1\ 0\ 0\ 0\ 0)) = 1,$$

so that $c = ab = (1\ 0\ 1\ 1\ 0\ 0\ 1)$.

D.4   Scalar Multiplication on Koblitz Curves

This section describes a particularly efficient method of computing the scalar multiple nP on the Koblitz curve E_a over $GF(2^m)$.

The operation τ is defined by:

$$\tau(x, y) = (x^2, y^2).$$

When the normal basis representation is used, then the operation τ is implemented by performing right circular shifts on the bit strings representing x and y.

Given m and a, define the following parameters:

- C is some integer greater than 5.

106

- $\mu = (-1)^{1-a}$.
- For $i = 0$ and $i = 1$, define the sequence $s_i(m)$ by:
$$s_i(0) = 0, \quad s_i(1) = 1 - i,$$
$$s_i(m) = \mu \bullet s_i(m-1) - 2\, s_i(m-2) + (-1)^i$$
- Define the sequence $V(m)$
$$V(0) = 2, \quad V(1) = \mu$$
$$V(m) = \mu \bullet v(m-1) - 2V(m-2).$$

For the example curves, the quantities $s_i(m)$ and $V(m)$ are as follows.

Curve K-163:

$s_0(163) =$ 2579386439110731650419537

$s_1(163) =$ –755360064476226375461594

$V(163) =$ –4845466632539410776804317

Curve K-233:

$s_0(233) =$ –27859711741434429761757834964435883

$s_1(233) =$ –44192136247082304936052160908934886

$V(233) =$ –137381546011108235394987299651366779

Curve K-283:

$s_0(283) =$ –665981532109049041108795536001591469280025

$s_1(283) =$ 1155860054909136775192281072591609913945968

$V(283) =$ 7777244870872830999287791970962823977569917

Curve K-409:

$s_0(409) =$ –18307510456002382137810317198756461378590542487556869338419259

$s_1(409) =$ –889304852613830409719665324184421267926566100996606444816790

$V(409) =$ 10457288737315625927447685387048320737638796957687575791173829

Curve K-571:

$s_0(571) =$ –373731944687646369242938589247611567147293964596131024123406420\
235241916729983261305

$s_1(571) =$ –3191857706446416099583814595948959674131968912148564658610565117\
58982848515832612248752

$V(571) =$ $-14838092698169141389961914029705149036454257418049393623291233951$
34208516828973111459843

The following algorithm computes the scalar multiple nP on the Koblitz curve E_a over $GF(2^m)$. The average number of elliptic additions and subtractions is at most $\sim 1 + (m/3)$, and is at most $\sim m/3$ with probability at least $1 - 2^{5-C}$.

1. For $i = 0$ to 1 do

   1.1 $n' \leftarrow \lfloor n / 2^{a-C+(m-9)/2} \rfloor$.

   1.2 $g' \leftarrow s_i(m) \cdot n'$.

   1.3 $h' \leftarrow \lfloor g' / 2^m \rfloor$.

   1.4 $j' \leftarrow V(m) \cdot h'$.

   1.5 $l' \leftarrow \text{Round}((g' + j') / 2^{(m+5)/2})$.

   1.6 $\lambda_i \leftarrow l' / 2^C$.

   1.7 $f_i \leftarrow \text{Round}(\lambda_i)$.

   1.8 $\eta_i \leftarrow \lambda_i - f_i$.

   1.9 $h_i \leftarrow 0$.

2. $\eta \leftarrow 2\eta_0 + \mu \eta_1$.

3. If $(\eta \geq 1)$,

   then

         if $(\eta_o - 3\mu\eta_1 < -1)$

             then set $h_1 \leftarrow \mu$

             else set $h_0 \leftarrow 1$.

   else

         if $(\eta_0 + 4\mu\eta_1 \geq 2)$

             then set $h_1 \leftarrow \mu$.

4. If $(\eta < -1)$

         then

       if $(\eta_0 - 3\mu\eta_1 \geq 1)$

           then set $h_1 \leftarrow -\mu$

           else set $h_0 \leftarrow -1$.

else

$$\text{if } (\eta_0 + 4\,\mu\,\eta_1 < -2)$$

$$\text{then set } h_1 \leftarrow -\mu.$$

5. $q_0 \leftarrow f_0 + h_0.$

6. $q_1 \leftarrow f_1 + h_1.$

7. $r_0 \leftarrow n - (s_0 + \mu\,s_1)\,q_0 - 2s_1\,q_1.$

8. $r_1 \leftarrow s_1\,q_0 - s_0\,q_1.$

9. Set $Q \leftarrow O.$

10. $P_0 \leftarrow P.$

11. While $((r_0 \neq 0) \text{ or } (r_1 \neq 0))$

    11.1 If $(r_0 \text{ odd})$, then

        11.1.1 set $u \leftarrow 2 - (r_0 - 2\,r_1 \bmod 4).$

        11.1.2 set $r_0 \leftarrow r_0 - u.$

        11.1.3 if $(u = 1)$, then set $Q \leftarrow Q + P_0.$

        11.1.4 if $(u = -1)$, then set $Q \leftarrow Q - P_0.$

    11.2 Set $P_0 \leftarrow \tau P_0.$

    11.3 Set $(r_0 , r_1) \leftarrow (r_1 + \mu r_0 /2, -r_0 /2).$

        Endwhile

12. Output $Q.$

D.5 Generation of Pseudo-Random Curves (Prime Case)

Let l be the bit length of p, and define

$$v = \lfloor (l-1)/160 \rfloor$$

$$w = l - 160v - 1.$$

1. Choose an arbitrary 160-bit string s as the domain parameter seed.

2. Compute $h = \text{SHA-1}(s).$

3. Let h_0 be the bit string obtained by taking the w rightmost bits of h.

4. Let z be the integer whose binary expansion is given by the 160-bit string s.

5. For i from 1 to v do:

5.1 Define the 160-bit string s_i to be binary expansion of the integer

$$(z + i) \bmod (2^{160}).$$

5.2 Compute $h_i = \text{SHA-1}(s_i)$.

6. Let h be the bit string obtained by the concatenation of h_0, h_1, ... , h_v as follows:

$$h = h_0 \parallel h_1 \parallel \ldots \parallel h_v.$$

7. Let c be the integer whose binary expansion is given by the bit string h.

8. If $((c = 0 \text{ or } 4c + 27 \equiv 0 \pmod{p}))$, then go to Step 1.

9. Choose integers $a, b \in GF(p)$ such that

$$c\, b^2 \equiv a^3 \pmod{p}.$$

(The simplest choice is $a = c$ and $b = c$. However, one may want to choose differently for performance reasons.)

10. Check that the elliptic curve E over $GF(p)$ given by $y^2 = x^3 + ax + b$ has suitable order. If not, go to Step 1.

D.6 Verification of Curve Pseudo-Randomness (Prime Case)

Given the 160-bit domain parameter seed value s, verify that the coefficient b was obtained from s via the cryptographic hash function SHA-1 as follows.

Let l be the bit length of p, and define

$$v = \lfloor (l - 1) / 160 \rfloor$$
$$w = l - 160v - 1.$$

1. Compute $h = \text{SHA-1}(s)$.

2. Let h_0 be the bit string obtained by taking the w rightmost bits of h.

3. Let z be the integer whose binary expansion is given by the 160-bit string s.

4. For $i = 1$ to v do

4.1 Define the 160-bit string s_i to be binary expansion of the integer

$$(z + i) \bmod (2^{160}).$$

4.2 Compute $h_i = \text{SHA-1}(s_i)$.

5. Let h be the bit string obtained by the concatenation of h_0, h_1, ... , h_v as follows:

$$h = h_0 \parallel h_1 \parallel \ldots \parallel h_v.$$

110

6. Let c be the integer whose binary expansion is given by the bit string h.

7. Verify that $b^2 c \equiv -27 \pmod{p}$.

D.7   Generation of Pseudo-Random Curves (Binary Case)

Let:

$$v = \lfloor (m - 1)/B \rfloor$$
$$w = m - Bv.$$

1. Choose an arbitrary 160-bit string s as the domain parameter seed.

2. Compute $h = $ SHA-1(s).

3. Let h_0 be the bit string obtained by taking the w rightmost bits of h.

4. Let z be the integer whose binary expansion is given by the 160-bit string s.

5. For i from 1 to v do:

    5.1   Define the 160-bit string s_i to be binary expansion of the integer
    $(z + i) \bmod (2^{160})$.

    5.2   Compute $h_i = $ SHA-1(s_i).

6. Let h be the bit string obtained by the concatenation of h_0, h_1, . . . , h_v as follows:

$$h = h_0 \parallel h_1 \parallel \ldots \parallel h_v.$$

7. Let b be the element of $GF(2^m)$ which binary expansion is given by the bit string h.

8. Choose an element a of $GF(2^m)$.

9. Check that the elliptic curve E over $GF(2^m)$ given by $y^2 + xy = x^3 + ax^2 + b$ has suitable order. If not, go to Step 1.

D.8   Verification of Curve Pseudo-Randomness (Binary Case)

Given the 160-bit domain parameter seed value s, verify that the coefficient b was obtained from s via the cryptographic hash function SHA-1 as follows.

Define

$$v = \lfloor (m - 1)/160 \rfloor$$
$$w = m - 160v$$

1. Compute $h = $ SHA-1(s).

2. Let h_0 be the bit string obtained by taking the w rightmost bits of h.

3. Let z be the integer whose binary expansion is given by the 160-bit string s.

4. For $i = 1$ to v do

    4.1 Define the 160-bit string s_i to be binary expansion of the integer $(z + i) \bmod (2^{160})$.

    4.2 Compute $h_i = \text{SHA-1}(s_i)$.

5. Let h be the bit string obtained by the concatenation of $h_0, h_1, \ldots, h_v$ as follows:

$$h = h_0 \parallel h_1 \parallel \ldots \parallel h_v.$$

6. Let c be the element of $GF(2^m)$ which is represented by the bit string h.

7. Verify that $c = b$.

D.9   Polynomial Basis to Normal Basis Conversion

Suppose that α is an element of the field $GF(2^m)$. Let p be the bit string representing α with respect to a given polynomial basis. It is desired to compute n, the bit string representing α with respect to a given normal basis. This is done via the matrix computation

$$p\,\Gamma = n,$$

where Γ is an m-by-m matrix with entries in $GF(2)$. The matrix Γ, which depends only on the bases, can be computed easily given its second-to-last row. The second-to-last row for each conversion is given the below.

Degree 163:

```
     3 e173bfaf 3a86434d 883a2918 a489ddbd 69fe84e1
```

Degree 233:

```
     0be 19b89595 28bbc490 038f4bc4 da8bdfc1 ca36bb05 853fd0ed
0ae200ce
```

Degree 283:

```
 3347f17 521fdabc 62ec1551 acf156fb 0bceb855 f174d4c1 7807511c
9f745382 add53bc3
```

Degree 409:

```
 0eb00f2 ea95fd6c 64024e7f 0b68b81f 5ff8a467 acc2b4c3 b9372843
6265c7ff a06d896c ae3a7e31 e295ec30 3eb9f769 de78bef5
```

Degree 571:

```
 7940ffa ef996513 4d59dcbf e5bf239b e4fe4b41 05959c5d 4d942ffd
46ea35f3 e3cdb0e1 04a2aa01 cef30a3a 49478011 196bfb43 c55091b6
1174d7c0 8d0cdd61 3bf6748a bad972a4
```

112

Given the second-to-last row r of Γ, the rest of the matrix is computed as follows. Let β be the element of $GF(2^m)$ whose representation with respect to the normal basis is r. Then the rows of Γ, from top to bottom, are the bit strings representing the elements

$$\beta^{m-1}, \beta^{m-2}, \ldots, \beta^2, \beta, 1$$

with respect to the normal basis. (Note that the element 1 is represented by the all-1 bit string.)

Alternatively, the matrix is the inverse of the matrix described in Appendix D.10.

More details of these computations can be found in Annex A.7 of the IEEE Standard 1363-2000 standard.

D.10 Normal Basis to Polynomial Basis Conversion

Suppose that α is an element of the field $GF(2^m)$. Let n be the bit string representing α with respect to a given normal basis. It is desired to compute p, the bit string representing α with respect to a given polynomial basis. This is done via the matrix computation

$$n\,\Gamma = p,$$

where Γ is an m-by-m matrix with entries in $GF(2)$. The matrix Γ, which depends only on the bases, can be computed easily given its top row. The top row for each conversion is given below.

Degree 163:

```
    7 15169c10 9c612e39 0d347c74 8342bcd3 b02a0bef
```

Degree 233:

```
    149 9e398ac5 d79e3685 59b35ca4 9bb7305d a6c0390b cf9e2300
253203c9
```

Degree 283:

```
 31e0ed7 91c3282d c5624a72 0818049d 053e8c7a b8663792 bc1d792e
ba9867fc 7b317a99
```

Degree 409:

```
 0dfa06b e206aa97 b7a41fff b9b0c55f 8f048062 fbe8381b 4248adf9
2912ccc8 e3f91a24 e1cfb395 0532b988 971c2304 2e85708d
```

Degree 571:

```
 452186b bf5840a0 bcf8c9f0 2a54efa0 4e813b43 c3d41496 06c4d27b
487bf107 393c8907 f79d9778 beb35ee8 7467d328 8274caeb da6ce05a
eb4ca5cf 3c3044bd 4372232f 2c1a27c4
```

Given the top row r of Γ, the rest of the matrix is computed as follows. Let β be the element of $GF(2^m)$ whose representation with respect to the polynomial basis is r. Then the rows of Γ, from top to bottom, are the bit strings representing the elements

$$\beta, \beta^2, \beta^{2^2}, \ldots, \beta^{2^{m-1}}$$

with respect to the polynomial basis.

Alternatively, the matrix is the inverse of the matrix described in Appendix D.9.

More details of these computations can be found in Annex A.7 of the IEEE Std 1363-2000 standard.

Appendix E: A Proof that $v = r$ in the DSA

(Informative)

The purpose of this appendix is to show that if $M' = M$, $r' = r$ and $s' = s$ in the signature verification, then $v = r'$. Let **Hash** be an approved hash function. The following result is needed.

Lemma: Let p and q be primes such that q divides $(p - 1)$, let h be a positive integer less than p, and let $g = (h^{(p-1)/q} \bmod p)$. Then $(g^q \bmod p) = 1$, and if $(m \bmod q) = (n \bmod q)$, then $(g^m \bmod p) = (g^n \bmod p)$.

Proof:

$$g^q \bmod p = (h^{(p-1)/q} \bmod p)^q \bmod p$$
$$= h^{(p-1)} \bmod p$$
$$= 1$$

by Fermat's Little Theorem. Now let $(m \bmod q) = (n \bmod q)$, i.e., $m = (n + kq)$ for some integer k. Then

$$g^m \bmod p = g^{n+kq} \bmod p$$
$$= (g^n \, g^{kq}) \bmod p$$
$$= ((g^n \bmod p)(g^q \bmod p)^k) \bmod p$$
$$= g^n \bmod p,$$

since $(g^q \bmod p) = 1$.

Proof of the main result:

Theorem: If $M' = M$, $r' = r$, and $s' = s$ in the signature verification, then $v = r'$.

Proof:

$$w = (s')^{-1} \bmod q = s^{-1} \bmod q$$
$$u1 = ((\textbf{Hash}(M'))w) \bmod q = ((\textbf{Hash}(M))w) \bmod q$$
$$u2 = ((r')w) \bmod q = (rw) \bmod q.$$

Now $y = (g^x \bmod p)$, so that by the lemma,

$$v = ((g^{u1} \, y^{u2}) \bmod p) \bmod q$$
$$= ((g^{\textbf{Hash}(M)w} \, y^{rw}) \bmod p) \bmod q$$
$$= ((g^{\textbf{Hash}(M)w} \, g^{xrw}) \bmod p) \bmod q$$
$$= ((g^{(\textbf{Hash}(M) + xr)w} \bmod p) \bmod q.$$

Also:

$$s = (k^{-1} (\mathbf{Hash}(M) + xr)) \bmod q.$$

Hence:

$$w = (k (\mathbf{Hash}(M) + xr)^{-1}) \bmod q$$

$$(\mathbf{Hash}(M) + xr)w \bmod q = k \bmod q.$$

Thus, by the lemma:

$$v = (g^k \bmod p) \bmod q = r$$

Appendix F: Calculating the Required Number of Rounds of Testing Using the Miller-Rabin Probabilistic Primality Test (Informative)

F.1 The Required Number of Rounds of the Miller-Rabin Primality Tests

The ideas of paper [1] were applied to estimate $p_{k,t}$, the probability that an odd k-bit integer that passes t rounds of Miller-Rabin (M-R) testing is actually composite. The probability $p_{k,t}$ is understood as the ratio of the number of odd composite numbers of a binary length k that can be expected to pass t rounds of M-R testing (with randomly generated bases) to the sum of that value and the number of odd prime integers of binary length k. This is equivalent to assuming that candidates selected for testing will be chosen uniformly at random from the entire set of odd k-bit integers. Following Pomerance, et al., $p_{k,t}$ can be (over) estimated by the ratio of the expected number of odd composite numbers of binary length k that will pass t rounds of M-R testing (with randomly generated bases) to the total number of odd primes of binary length k. From the perspective of a party charged with the responsibility of generating a k-bit prime, the objective is to determine a value of t such that $p_{k,t}$ is no greater than an acceptably small target value p_{target}.

Using [1], it is possible to compute an upper bound for $p_{k,t}$ as a function of k and t. From this, an upper bound can be computed for t as a function of k and p_{target}, the maximum allowed probability of accidentally generating a composite number. The following is an algorithm for computing t:

1. For $t = 1, 2 \ldots \lceil -\log_2(p_{target})/2 \rceil$

    1.1 For $M = 3, 4 \ldots \lfloor 2\sqrt{k-1} - 1 \rfloor$ (1)

        1.1.1   Compute $p_{k,t}$ as in (2).

        1.1.2   If $p_{k,t} \leq p_{target}$

              1.1.2.1   Accept t.

              1.1.2.2   Stop.

In (1), k is the bit length of the candidate primes and (2) is as follows:

$$p_{k,t} = 2.00743 \cdot \ln(2) \cdot k \cdot 2^{-k} \left[2^{k-2-Mt} + \frac{8(\pi^2 - 6)}{3} 2^{k-2} \sum_{m=3}^{M} 2^{m-(m-1)t} \sum_{j=2}^{m} \frac{1}{2^{\left(j + \frac{(k-1)}{j} \right)}} \right] . \qquad (2)$$

Using this expression for t, the following methodologies are used for testing the DSA and RSA candidate primes.

F.2   Generating DSA Primes

For DSA, the maximum possible care must be taken when generating the primes p and q that are used for the domain parameters. The same primes p and q are used by many parties. This means that any weakness that these numbers may possess would affect multiple users. It also means that the primes are not generated very often; typically, an entire system uses the same set of domain parameters for an extended period of time. Therefore, in this case, some additional care is called for.

With this in mind, it may be too optimistic to simply subject candidate primes to t rounds of M-R testing, where the minimal acceptable value for t is determined according to (1) and (2) in Appendix F.1. This might be the case, for example, if there is a reason to doubt that the assumptions made in [1] have been satisfied during the process of selecting candidates for primality testing. One may gain more confidence in the process by performing some additional (different) primality test(s) on the candidates that survive the M-R testing. As another option, one could, of course, perform additional rounds of M-R testing. These considerations lead to the following alternatives: either (A) use the number of rounds of M-R testing determined according to (1) and (2) in Appendix F.1, and follow that with a single Lucas test (as recommended in ANS X9.31), or (B) use a (much) more conservative approach when determining t (e.g., as described below) and subject candidate primes to additional rounds of M-R testing.

One approach for strategy (B) would be to adopt the viewpoint of the majority of system users, who have no part in generating the (supposed) prime, but who must rely upon its primality for their security. Such parties may be concerned that the candidates for M-R testing have been selected in a fashion that deviates significantly from the uniform distribution – which was assumed when determining t according to (1) and (2) in Appendix F.1. In cases where the selection process could be unusually biased in some way, it is important to minimize the probability that a composite number will survive testing. It can be shown that for any k-bit odd composite number (regardless of how it was selected), the probability that it will pass t rounds of M-R testing with randomly chosen bases is less than 4^{-t} (although this is not a particularly tight bound). Selecting t such that $4^{-t} \leq p_{target}$ is equivalent to choosing $t \geq -\log_2(p_{target})/2$. To ensure that a composite number has a probability no greater than p_{target} of surviving the M-R tests, the number of rounds can be set at $t = \lceil -\log_2(p_{target})/2 \rceil$. Even if the method of selecting candidates were so biased that it offered nothing but composite numbers for testing, it is reasonable to expect that it would take at least $1/p_{target}$ attempts (which is greater than 4^t) before a composite number would slip through the t-round M-R testing process.

WARNING: As the discussion above illustrates, care must be taken when using the phrase "error probability" in connection with the recommended number of rounds of M-R testing. The probability that a composite number survives t rounds of Miller-Rabin testing is <u>not</u> the same as

$p_{k,t}$, which is the probability that a number surviving t rounds of Miller-Rabin testing is composite. Ordinarily, the latter probability is the one that should be of most interest to a party responsible for generating primes, while the former may be more important to a party responsible for validating the primality of a number generated by someone else. However, for sufficiently large k (e.g., $k \geq 51$), it can be shown that $p_{k,t} \leq 4^{-t}$ under the same assumptions concerning the selection of candidates as those made to obtain formula (2) in Appendix F.1 (see [1].) In such cases, $t = \lceil -\log_2(p_{target})/2 \rceil$ rounds of Miller-Rabin testing can be used both in generating and validating primes, with p_{target} serving as an upper bound on both the probability that the generation process yields a composite number and the probability that a composite number would survive an attempt to validate its primality.

Table C.1 in Appendix C.3 identifies the minimum values for t when generating the primes p and q for DSA using either strategy (A) or (B) above. To obtain the t values shown in the column titled "M-R Tests Only", the conservative strategy (B) was followed; those t values are sufficient to validate the primality of p and q. The t values shown in the column titled "M-R Tests when followed by One Lucas Test" result from following strategy (A) using computations (1) and (2) in Appendix F.1.

F.3   Generating Primes for RSA Signatures

When generating primes for the RSA signature algorithm, it is still very important to reduce the probability of errors in the M-R testing procedure. However, since the (probable) primes are used to generate a user's key pair, if a composite number survives the testing process, the consequences of the error may be less dramatic than in the case of generating DSA domain parameters; only one user's transactions are affected, rather than a domain of users. Furthermore, if the p or q value generated for some user is composite, the problem will not be undiscovered for long, since it is almost certain that signatures generated by that user will not be verifiable.

Therefore, when generating the RSA primes p and q, it is sufficient to use the number of rounds derived from (1) and (2) in Appendix F.1 as the minimum number of M-R tests to be performed. However, if the definition of $p_{k,t}$ is not considered to be sufficiently conservative when testing p and q, it is recommended that the t rounds of Miller-Rabin tests be followed by a single Lucas test.

The lengths for p and q that are recommended for use in RSA signature algorithms are 512, 1024 and 1536 bits; recall that $n = pq$, so the corresponding lengths for n are 1024, 2048 and 3072 bits, respectively. As currently specified in SP 800-57, Part 1, these lengths correspond to security strengths of 80, 112 and 128 bits, respectively. Hence, it makes sense to match the number of rounds of Miller-Rabin testing to the target error probability values of 2^{-80}, 2^{-112}, and 2^{-128}. A probability of 2^{-100} is included for all prime lengths, since this probability has often been used in the past and may be acceptable for many applications.

When generating the RSA primes p and q with conditions, it is sufficient to use the value t derived from (1) and (2) as the minimum number of M-R tests to be performed when generating the auxiliary primes p_1, p_2, q_1 and q_2. It is not necessary to use an additional Lucas test on these numbers. In the extremely unlikely event that one of the numbers p_1, p_2, q_1 or q_2 is composite, there is still a high probability that the corresponding RSA prime (p or q) will satisfy the requisite conditions.

The sizes of p_1, p_2, q_1, and q_2 were chosen to ensure that, for an adversary with significant but not overwhelming resources, Lenstra's elliptic curve factoring method [2] (against which there is no protection beyond choosing large p and q) is a more effective factoring algorithm than either the Pollard P–1 method [2], the Williams P+1 method [3] or various cycling methods [2]. For an adversary with overwhelming resources, the best all-purpose factoring algorithm is assumed to be the General Number Field Sieve [2].

Tables C.2 and C.3 in Appendix C.3 specify the minimum number of rounds of M-R testing when generating primes to be used in the construction of RSA signature key pairs.

Appendix G: References

[1] I. Damgard, P. Landrock, and C. Pomerance, C. "Average Case Error Estimates for the Strong Probable Prime Test," Mathematics of Computation, v. 61, No, 203, pp. 177-194, 1993.

[2] A.J Menezes, P.C. Oorschot, and S.A. Vanstone. Handbook of Applied Cryptography. CRC Press, 1996.

[3] H.C. Williams. "A p+1 Method of factoring". *Math. Comp.* 39, 225-234, 1982.

[4] D.E. Knuth, The Art of Computer Programming, Vol. 2, 3rd Ed., Addison-Wesley, 1998, Algorithm P, page 395.

[5] R. Baillie and S.S. Wagstaff Jr., Mathematics of Computation, V. 35 (1980), pages 1391 – 1417.